Efficient Livestock Handling

Efficient Livestock Handling

The Practical Application of Animal Welfare and Behavioral Science

Bonnie V. Beaver
College of Veterinary Medicine, Texas A&M University,
College Station, TX, USA

Donald L. Höglund
Veterinarian and Educator, Livestock Trust Institute,
Raleigh, North Carolina, USA

AMSTERDAM • BOSTON • HEIDELBERG • LONDON
NEW YORK • OXFORD • PARIS • SAN DIEGO
SAN FRANCISCO • SINGAPORE • SYDNEY • TOKYO

Academic Press is an imprint of Elsevier

Academic Press is an imprint of Elsevier
125 London Wall, London EC2Y 5AS, UK
525 B Street, Suite 1800, San Diego, CA 92101-4495, USA
225 Wyman Street, Waltham, MA 02451, USA
The Boulevard, Langford Lane, Kidlington, Oxford OX5 1GB, UK

ISBN: 978-0-12-418670-5

British Library Cataloguing-in-Publication Data
A catalogue record for this book is available from the British Library

Library of Congress Cataloging-in-Publication Data
A catalog record for this book is available from the Library of Congress

For information on all Academic Press publications
visit our website at http://store.elsevier.com/

Working together
to grow libraries in
developing countries

www.elsevier.com • www.bookaid.org

Publisher: Sarah Tenney
Senior Acquisition Editor: Kristi A.S. Gomez
Senior Editorial Project Manager: Pat Gonzalez
Production Project Manager: Lucía Pérez
Designer: Matthew Limbert

Typeset by TNQ Books and Journals
www.tnq.co.in

Dedication

We are pleased to dedicate this book to people who care about providing the best welfare to the animals they work with by using practices based on sound science. In addition, each of us would like to dedicate this to some special individuals. For Don, that is Dr Karla Andrews, his wife. For Bonnie, the dedication goes to the Charter Diplomates of the American College of Animal Welfare, who shared the task of creating this veterinary specialty organization.

Contents

About the Authors

Bonnie V. Beaver

Dr Beaver is a veterinarian internationally recognized for her work in animal welfare and in the normal and abnormal behaviors of animals. She has given over 250 scientific presentations on these subjects and has discussed many areas of veterinary medicine for the public media. In addition, she has authored over 150 scientific articles and has nine published books.

Dr Beaver is a member of numerous local, state, and national professional organizations and has served as a Charter Diplomate and president of the American College of Veterinary Behaviorists and the American College of Animal Welfare. She has also served as president of the Texas Veterinary Medical Association and the American Veterinary Medical Association.

Donald L. Höglund

Dr Höglund is a veterinarian internationally recognized for his work with humane handling techniques in livestock. As CEO of the Livestock Trust Institute, he has focused his career on educating and training stock handlers on safe, efficient, and humane ways to work with animals in agriculture. His early work helped rehabilitate 4000 prison inmates while handling more than 20,000 free-roaming mustangs.

His best-selling book, *Nobody's Horses*, told the true story of the capture and rescue of 2000 former ranch horses from the White Sands Missile Range, the birthplace of the world's first atomic detonation.

Twice acknowledged for humanitarian effort by the United States Congress, Dr Höglund also played a central production role in the establishment of "Buffalo Bill's Wild West" at Disneyland-Paris.

Introduction

Humane handling practices are not just about the animal. They are also about the people who interact with the animals. Animals learn 24 hours a day, 7 days a week, and 365 days a year. That means that every encounter between animals and people can shape the future behavior of both. Livestock observe the world around them and respond based on previous experiences, sensory input, inherited traits, and species-typical behaviors. In that way, they are no different than people. There are differences, however. These lie in different experiences, perceptions, and behavior.

Knowledge of animal behavior and neurology have matured to the point that veterinarians and animal scientists can now use terms and phrases that respect the differences between animals and humans. Humans have the unique potential for verbal language, and that underlies some of the fundamental distinctions between the species. The belief that language, sensory perception, cognition, and culture shape human experiences, including emotions, is well accepted in neuroscience.[1–3] The lack of language and culture, and the differences in sensory capabilities mean that animals cannot experience the world the way humans do. That also means that animal feelings, whatever they might be, cannot be the same.[4]

That is not a denial of conscious awareness in animals; instead, it is recognition that humans will never truly understand their perspective. Scientists can study whether neurological and physiological responses to situations parallel those of humans. While sentience is widely accepted for mammals, we do not know if animals interpret senses the same way a person might. Regardless of the variations between livestock and humans or in the similarities of overt expressions of behavior, gentle, consistent, and nonthreatening handling will help to establish a relationship that empowers safe, efficient, humane, and productive results for animals and their human handlers.

Humane handling is important, but what role does it play in the practical welfare debate? The most obvious reason is that inappropriate handling results in conflict behaviors potentially leading to stress in animals and handlers. But what exactly is stress, or worse, distress? We use these terms often and try to measure them physiologically, but scientists do not have a universally accepted definition.[5] That reality makes stress difficult to study, qualify, and quantify. The dilemma in using the word stress is enhanced because subjective evaluations of "stress" are individual assessments. That subjectivity can result in endless opposing interpretations or value assumptions.[6,7] Practically, though, it is easiest to make the assumption that something that is painful or stressful to humans will also be painful or stressful to animals.

Stress is perceived differently by individuals and is not necessarily bad. Although the word stress is often used to mean distress, they really are different. Distress is more extreme and is said to occur when the animal cannot adapt to negative internal or external conditions.[8] As research defines more accurate ways of measuring the stress response, animal handlers will be better able to identify and reduce the incidence or severity of stressors. That has positive welfare implications for all animals.

Assessments of humane handling are based on several factors, including the use of techniques that minimize suffering and promote contentment; allow good health, growth, and reproduction; or allow the animal to use natural behaviors.[7]

In recent years, there has been an increased emphasis on techniques that improve the welfare of all species of animals. Improved handling techniques are only part of the solution. Understanding how animals learn is arguably the cornerstone for studying, teaching, and applying efficient handling methods wherever animals reside. Coupling the science of learning with practical applications is where the art and science of humane animal handling converge. Studied practice is likely the only way to perfect safe, humane, and efficient handling skill.

At the extreme, handling can be dangerous if the conflicted animal turns aggressive because it cannot find a route to escape distressful situations. This is the blatantly obvious concern. The really significant ones are more subtle. Studies in livestock behavior have repeatedly shown that aversive handling will increase maladaptive behaviors toward humans, making the animal's behavior unpredictable and potentially dangerous. In the final analysis, handler-created conflict behaviors in livestock can reduce production efficiency, growth, reproduction, and health.[9,10]

The 2011 National Beef Quality Audit found that 93% of US producers believe they use good stockmanship practices.[11] To some extent, this was supported by data showing a reduction in the number of bruises on cattle entering the plants over the previous 5 years, and an indication by producers that 98% do not use an electric prod as their primary driving tool.[11] Eliminating the use of the electric prod is a huge step toward humane handling because it is often used inappropriately; however, the words "as their primary diving tool" can really mean the range of use is from "not used at all" to "if they don't take a step right away I'll zap them." The difficulty is that objective evaluations are not always possible, particularly when there are not sufficient criteria to make an unbiased judgment. There can also be conflicting interpretations or value assumptions.[6,7]

Most of the humane techniques that have been popularized recently emphasize the use of the natural responses of an animal species to make working with them easier. Different authors have slightly different approaches to their techniques, but all are based on the concept of using what is natural for the animal: for example, moving away from an approaching alien species. Natural, instinctive behaviors and reflexes, however, are not a product of learning, and learning is the central component of the humane handling and training of animals. Without an understanding of how animals interpret their environment and learn from it, handlers are merely doing things—standing here and moving there. Understanding what natural survival behaviors are and how they can be incorporated to influence learning can help to make effective handling efficient. If the handling process is humane and efficient, then it is effective and productive too.

In any environment, it is desirable for people working with animals to have a variety of humane handling techniques to choose from. Handlers need to adapt what they try to what works efficiently and to which of the humane methods feel the most comfortable to them. Each animal is one-of-a-kind, and every handling scenario is unique from the animal's perspective because it ultimately needs to adapt what is tried to what feels comfortable.

We do not have data regarding the handling of livestock species other than that mentioned above, and there certainly has been a large amount of negative press regarding some high visibility handling situations spread by the media. Regardless, most people think they do a good job at using humane handling techniques, even if they truly are not. Attitudes of stockmen and women regarding aversive handling of their livestock have been shown to be independent of their attitudes toward the animals, even though they recognize that human contact is important to the ease of handling.[12]

Adopting new handling styles is difficult once a pattern has been learned. That is true for animals, and it is especially true for people. It is easier to buy a new piece of handling equipment than it is to adopt a new technique, even one that is easy to learn.[13] It sometimes takes an outside influence to make us take a fresh look at what we are actually doing. As an example, a veterinary colleague was teaching students how to dehorn a calf using the typical method of physical restraint while cutting the horn bud. Some of the students who had not been raised with cattle asked why ring-block anesthesia was not used. The veterinarian used the ring-block on the next calf and was totally amazed at how calmly the calf responded. The willingness to try something different led to a humane discovery for this veterinarian that was less painful for the calf and safer for the people who were handling the animal. The other invisible benefit was that the calf did not learn to avoid people.

To help readers make the transitions, we have included general information about how the various livestock species perceive the world, learn from it, and react to it. Understanding these things will help make sense of the procedures. Replacing old practices with new techniques is difficult enough. Trying to complete the technique efficiently when the environment is not conducive to success only adds to the problem. Timing can be everything, but place, space, competing stimuli, and patience are also variables affecting success. Loud humans, barking dogs, popping whips, instant experts, limited time, and speed training position every scenario for failure.

Pictures are helpful to illustrate the descriptions for a method or technique; however, the ability to access videos that demonstrate the handler starting position, angle of animal approach, and the rate and intensity of approach or withdrawal is now possible. When a procedure makes sense, it is more likely to be tried successfully.

The contents of the book have been arranged to present both general information about things that are common to horses, cattle, pigs, sheep, and goats, and then provide variations specific to each of the species. Historical information allows the reader to understand how the various animals became associated with humans and how that relationship has changed to the point where "humane" is being incorporated into their interspecies interactions. From there, the reader will be able to gain insight into specific sensory capabilities and specific behaviors that influence animal reactions to handling. These basic areas can be modified by genetic, hormonal, and environmental factors, with many of these factors under human control to some degree. Knowledge of the principles of how animals learn is critical to understanding how to apply them to humane handling for the most efficient applications of the handling techniques of various livestock species within today's environment.

The book is not intended to cover specific behavior problems. It is our hope that readers develop a better understanding of the animals they work with, humane methods for animal handling, and techniques that suppress the development of behavior problems. We link the science of behavior and welfare with their practical applications. The intention is to improve the quality of life for all livestock and the handlers that serve them.

REFERENCES

1. Bowerman M, Levinson SC, editors. *Language acquisition and conceptual development*. New York: Cambridge University Press; 2001. p. 602.
2. Gentner D, Goldin-Meadow S, editors. *Language in mind: advances in the study of language and thought*. Cambridge (MA): MIT Press; 2003. p. 528.
3. Whorf BL. *Language, thought, and reality*. Cambridge (MA): Technology Press of MIT; 1956. p. 304.
4. LeDoux JE. Coming to terms with fear. *Proc Natl Acad Sci* February 25, 2014;**111**(8):2871–8.
5. Sapolsky RM. Endrocrinology of the stress-response. In: Becker JB, Breedlove M, Crews D, McCarty MM, editors. *Behavioral endocrinology*. 2nd ed. Cambridge (MA): MIT Press; 2002. p. 574.
6. Broom DM, Fraser AF. *Domestic animal behaviour and welfare*. 4th ed. Cambridge (MA): CAB International; 2007. p. 438.
7. Fraser D. Assessing animal welfare at the farm and group level: the interplay of science and values. *Anim Welfare* November 2003;**12**(4):433–43.
8. Rajeshwari YB. *Handbook on care and management of laboratory animals and pet animals*. New Delhi: New India Publishing; 2009. p. 302.
9. Hemsworth PH. Human-animal interactions in livestock production. *Appl Anim Behav Sci* May 2, 2003;**81**(3):185–98.
10. Hemsworth PH, Barnett JL, Beveridge L, Matthews LR. The welfare of extensively managed dairy cattle: a review. *Appl Anim Behav Sci* February 1995;**42**(3):161–82.
11. 2011 National Beef Quality Audit. Available at: http://www.bqa.org/CMDocs/bqa/NBQA_Significant_Findings.pdf; [accessed 03.01.2013].
12. Boivin X, Marcantognini L, Boulesteix P, Godet J, Brulé A, Veissier I. Attitudes of farmers towards Limousin cattle and their handling. *Anim Welfare* May 2007;**16**(2):147–51.
13. Grandin T. Transferring results of behavioral research to industry to improve animal welfare on the farm, ranch and the slaughter plant. *Appl Anim Behav Sci* May 2003;**81**(3):215–28.

Chapter 1

Historical Perspectives of Livestock Handling

I have been studying the traits and dispositions of the 'lower animals' (so called) and contrasting them with the traits and dispositions of man. I find the result humiliating to me.

Mark Twain

Doc Smith had been a veterinarian in East Texas approaching 50 years and he had about seen it all. He grew up on a cattle ranch that bordered the Trinity River. That meant he was a pretty good hand with a horse too. Today, Doc was leaning against the crumpled remains of a livestock panel, thinking about all the chaos he had just witnessed, when the local neighborhood "cow whisperer" was attempting to separate and load a couple of first calf heifers into his rickety old trailer.

Mr Cow Whisperer was talking the talk about the right way to get this task done as he swiped at the sweat trickling from under his Gator roping cap. It had been a difficult lesson, but "by-golly" they were now trapped between two cattle panels held together by bright orange baling twine and the rusty hinged back gate of the ramshackle stock trailer. Five of the nearby neighbors were drinking a beer as they loitered around an old tree, exhausted and sweaty from the cattle hustling events of that early morning fiasco.

Doc had about all he could stand. He knew too many who could talk the talk but not walk the walk. He kept reminding himself that handling livestock is about safe, efficient use of human presence and action, not about testosterone-driven human males who love to chase cattle. The veterinarian waved the men back from the startled heifers and began a pointed discussion about how humans can affect the cattle's reactions. It was a wonderful opportunity to discuss the benefits of low-distress stock handling.

Herding cattle has been a pastoral occupation since cattle were first domesticated. The lessons from this morning's cattle drive, or better termed cattle chasing, served to teach the animals to run from humans, noisy dirt bikes, and rapacious four-wheelers. The wild precursors of today's livestock needed gentle handling that took advantage of their responses to humans. The animals were too big and too leery to use "bruteacaine." The relationship between animal and

Efficient Livestock Handling. http://dx.doi.org/10.1016/B978-0-12-418670-5.00001-9

human has changed over the years as has the animal itself. Finally, people like Doc are coming back to the knowledge of animal behavior and the principles of stockmanship to safely, effectively, and efficiently work with livestock. This tale goes back to the ancestors of today's domesticated species because much of the genetic makeup of physical and behavioral traits comes from them.

"Humane handling" of livestock is a popular buzz word within the livestock industry, which to some extent has been driven by the general public. One of the problems with the use of the term is that interpretations of what it actually means vary considerably. There are also many techniques that can accomplish that same thing. The better stockmen and women understand the animals they work with, the easier it is to include humane handling techniques. That said, understanding of animals begins long ago, with domestication. That is the beginning of agriculture, and that is when the profound change in the relationship between humans and animals also started.[35]

DOMESTICATION

The domestication of today's livestock species occurred gradually over thousands of years, involving several processes and domestication events in diverse locations.[4] The main criterion connected with the start of the domestication process was a human need for something that a particular species could bring to human survival. The animals with undesirable traits were culled over time, so that the animals remaining were more likely to pass on the genes that resulted in usable features. Initially, the uses were diverse—dung for fires, meat and milk for nutrition, strength for transportation, and hides for clothing and housing. In different parts of the world, each species might have had a different role, and that role often changed over time.

The process of domestication involves more than simply taming individual animals. It involves a core group of wild animals that are separated from their wild herd mates and selectively bred over several generations to concentrate desirable traits, as well as their ability to be controlled by humans. While selective breeding can be initiated simply because the tamer individuals tend to hang around human habitation, which happened during dog domestication, it can also occur because humans force the separation, which happened with our livestock species.[10,14]

The end product, as we know it, is a species that has been bred in captivity over many generations by a human community that maintains total control over its breeding, organization of territory, and food supply.[2,5,6,10] It can actually be argued that domestication is not static and that it continues to be a dynamic, locally based, and ongoing process.[14]

Certain characteristics in wild animals ease the domestication process.[7] Group structure is the first category related to domestication that applies to all of today's livestock species and their wild ancestors. Animals that live in large, hierarchical structured groups are favored. This allows humans to keep large numbers in minimal spaces without the concern for territorial and resource disputes that might result in injuries. Promiscuous mating behaviors allow humans to select mates based on traits of our choosing. Multiple estrous cycles provide several opportunities to breed each year, and clear sexual readiness signals make it easy to detect when breeding will need to occur. Species that are not particularly wary of human presence also aid in ultimate domestication. Other characteristics that make a species more easily domesticated include omnivorous diets (rather than specialized ones) and diets that are by-products of (or not in competition with) human food production. Wild species, which are adaptable to a wide range of environments and have limited agility, also make domestication easier.

Not every species that has desirable traits is a candidate for domestication.[9] For example, wildebeest are close relatives of cattle and could be a source of meat. While they live in large groups like cattle, their migratory behaviors make them undesirable to try to domesticate. Wild pigs, on the other hand, have most of the characteristics that favored domestication of some in the past. When we think about the characteristics of the wild ancestors of modern livestock species, it makes sense why each species was able to undergo domestication.

The result of selective breeding over time is a domesticated animal that possesses heritable characteristics that distinguish it from its wild progenitors. Physical features are the most obvious of the three types of changes that tend to occur over time. The wild auroch was large compared to modern cattle. That is also true for the ancestors of sheep and goats. Wild hogs have a pointed snout and considerable body hair, while domesticated ones do not. Physiological changes, the second type of change, are less obvious until you consider that wild animals are stressed being around humans, while domestic ones are easily tamed to allow people to calmly interact directly with them. Overall, stress responses are geared down compared to those of their wild cousins. Dairy cattle provide another example of how a physiological modification of milk production has changed from enough to feed a single calf to way in excess of that. The number of piglets born to a sow has increased significantly too. The third type of change that comes with domestication is behavioral. Some behaviors are similar to those of the wild relatives, some have been altered slightly, and still others have been added or deleted. Although horses will still try to run from things that scare them, a behavior that is problematic when kept in fenced pastures, they can be genetically programmed to follow and work cattle or move with specialized gaits.

Because domestication occurs over several animal generations, it is not possible to pinpoint a specific year when domestication occurred for each species. This is compounded by the likelihood that some species had multiple domestication sites because of the wide distribution of the wild relatives. Recent studies using mitochondrial (mt) DNA (mtDNA) have revealed a large variation and complexity in domestication in terms of numbers and types of progenitors and with the number of times similar stocks were domesticated. Interestingly, most of the livestock species we have today descended from species that were

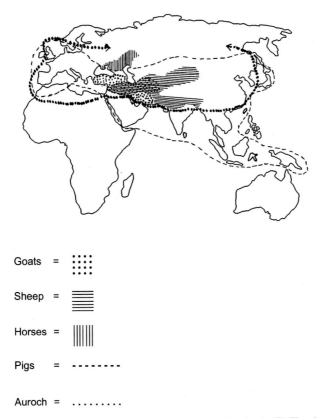

Goats = ::::

Sheep = ≡

Horses = ||||||

Pigs = - - - - - - - -

Auroch = ...

FIGURE 1.1 Distribution of progenitors of domestic animals, showing the likelihood of domestication of livestock species occurring in Western Asia.

concentrated in Western Asia, mainly near the Fertile Crescent of Mesopotamia, making it the location for most of the domestication events (Figure 1.1). There are some exceptions in Central and Eastern Asia.

Sheep and Goats

Sheep and goats were probably the first of the livestock species to be domesticated at least 9000, and perhaps as much as 10,500, years ago.[6,7,12,42] Modern sheep (*Ovis aries*) arose from the wild Asiatic mouflon (*Ovis orientalis*).[5,6,20] It is likely there were at least two separate domestication events, which correlates somewhat with the modern fat- and thin-tailed phenotypes.[4,14,20] Sheep were valuable to early people because they could be tended in one location or herded with nomads, existing wherever grazing was possible. Goats filled the higher altitude browsing niche, and the domesticated version (*Capra hircus*) originated from the Bezoar goat (*Capra aegagrus*), primarily in what is now Turkey.[5,6,30] For this species too, mtDNA shows at least two domestication events, one in the Near East and one in Asia.[4,12] Although goats are relatively closely related to sheep, they are behaviorally divergent and require different herding tactics.[14] They are able to survive in more marginal conditions and still provide most everything needed for human survival. As a result of being hardy and easy to move with migrants, goat mtDNA is very similar around the world (only 10% difference), especially when compared to cattle (greater than 50% variation) and sheep (35% variation).[4,14]

For both sheep and goats, their compact size and their ability to be herded made domestication relatively easy.[34] Features that we now associate with domestic versions might have actually come about by what has been called "unconscious selection."[14,43] As an example, the horns on wild sheep and goats take a lot of energy to produce but are needed to fend off reproductive rivals. Since fighting is not necessary for reproductive success in a managed flock/ herd, a mutation for smaller horns would allow a diversion of energy toward something else, such as earlier sexual maturity or higher sperm quality.[14]

Cattle

The wild auroch had a wide range of existence, and subspecies of it gave rise to modern cattle. *Bos primigenius primigenius* is the progenitor of the domestic humpless cattle (*Bos taurus*), and modern humped cattle (*Bos indicus*) descended from an extinct Indian auroch (*Bos primigenius namadicus*).[4-6,29] The two primitive subspecies of *Bos primigenius* originally diverged about 2 million years ago.[19] While most sources suggest the two modern types of cattle were domesticated about 8000 years ago, recent work suggests it may have been 2000–2500 years before that.[3] Regardless, the DNA profiling has revealed some interesting finds. The humped cattle of Africa show mtDNA that is very close to that of the humpless cattle and quite distant from the Indian humped ones. Evaluation of Y chromosomal patterns in males suggests that the hump came from bulls that were imported into Africa.[4,26] As the humpless cattle migrated with humans northwestward into Europe, it is likely that they interbred with local auroch bulls, as indicated in the analysis of the Y chromosome haplotype in modern European cattle breeds.[16] Recent research shows that the Texas Longhorn cattle are most closely related to African cattle.[29] It is speculated that the Moors brought their cattle to the Iberian Peninsula, and descendants of those cattle crossed the Atlantic with Columbus' second expedition to the New World. Escapees had inherited some of the hardiness needed for harsh environments from the humped cattle many years earlier. Over time as feral animals, these cattle developed the large horn size in response to evolutionary pressures for survival.

There is a debate as to why humans would even try to interact with a wild species that stood about five feet at the shoulder. It has been argued that for ancient people, auroch bulls were important symbols, and domestication may have been for religiopolitical reasons.[34] It is known that in ancient times, cattle were not kept for meat or milk. It might have been for purposes of sacrifice and barter instead.[5] Researchers speculate that the domestication of cattle was accompanied by castration, nose-ringing, and tipping of horns so that humans could more safely work with such powerful animals.[34]

Pigs

Wild boars (*Sus scrofa*) had the widest distribution of all the early species, extending from Europe into what is now Southeast Asia. The ancestral forms of the European and Asian subspecies are estimated to have diverged from each other between 500,000 and one million years ago.[15,17] Then around 8000 years ago, modern pigs (*Sus domesticus*) were domesticated in multiple sites. The mtDNA evidence suggests that there were at least six independent domestication events, ranging from one involving the Near East/European wild boar species to one from an Eastern Asian subspecies.[4,14,23,24,41] Initially, pig meat was not particularly favored, compared to that from sheep and goats, but as "farming" moved from Western Asia into Northern and Western Europe, cattle and pigs were more highly favored.[4,5] Inbreeding between domestic and wild boar species probably happened numerous times, particularly in Europe, where DNA traces of Near Eastern domestic pigs were eventually replaced almost entirely by domesticated local stock.[22] Continued strong selection within the European pigs gave rise to anatomical differences that now distinguish them, including the elongation of the back and an increased number of vertebrae.[32] Even with these significant anatomic changes and years of selection toward domestication, only 7% of the genome has been changed.[1]

Horses

The domestication of horses changed human history. Horses were the last of the livestock species to be domesticated, about 5000–6000 years ago, and the center for this was located in several sites closer to Central Asia in the general area of what is now Ukraine.[7,33] The mtDNA suggests that there were at least seven distinct populations of wild horses (*Equus ferus*) that were domesticated into the modern horse (*Equus caballus*).[18] In addition to the DNA evidence, coat color variations also began to show up about the same time, suggesting that selective breeding had started to occur.[25] As horses expanded from Central Asia, it is likely that they occasionally bred with local wild horses.[33,39,40] As the numbers of domesticated horses grew, the uses also expanded. They supplied meat, milk, and transportation.[28,39] While peasants continued to use donkeys, which had been domesticated slightly earlier, the rich and powerful quickly replaced their donkeys with horses. The speed and power of the horse meant nobility could now extend their reach with horse-drawn war chariots.[6,34] DNA characteristics are very similar in all horses, indicating that their use for transportation was widespread.[4]

HUMAN–ANIMAL RELATIONSHIPS

The relationship between humans and animals has been complex and certainly has varied over time both by individuals and societies. People who work with livestock are very much aware of recent societal pressures, and understanding what these pressures are helps put the concepts of how animals are handled into perspective. The specific events that drove the domestication of animals in the Fertile Crescent are not known; however, the climate in that area of the world had been favorable to a shift from hunting to agriculture, and it is speculated that the human population was expanding as well. Hunting allowed nomads to kill what was needed for sustenance, but farming could produce more than enough to feed the population. The downside to farming is that producers needed some way to preserve food stores. In the hot and semiarid environments, this could best be completed with grain and live animals. And animals that looked different than their wild relatives would help farmers keep the latter away from crops, making selective breeding for physical traits helpful.[7]

Pastoralism is one of the oldest views of animals and dates back to the start of domestication. More recently, we have come to associate it with biblical origins.[13] Interpretations of the Bible vary from one of "caretaker" to one of "dominance over." In that later view, humans had souls and the ability to think and reason, but animals did not. So, humans were considered to be superior and needed to care for "inferior" beings.[8,36] As long as certain conventions were used, such as a day of rest and ritualized slaughter, animals were allowed to fulfill certain purposes, such as for meat, transportation, and clothing. Rescuing an animal was one of the few things permitted on the Sabbath. The early church did differentiate between animal species and allowed "justice" toward noxious beasts.[8,27]

The strong attachment with animals was evidenced in ancient peoples, even in the mythical sense. Auroch bulls, and cows by association, were revered because of their size, strength, and fertility. They were worshipped or at least held in very high regard in a number of different cultures. The ancient Greeks called bulls *itali*, which ultimately gave rise to *Italy*, to honor the number and beauty of its cattle. Humans named heavenly constellations in honor of bulls or their parts. Wealth was measured in terms of the number of cattle a person had and eventually led to pictures of cattle on coins. Even today, there are remnants of the relationships: cattle remain sacred in Hinduism, and the word *chattel* meaning "movable property," as examples. On the other hand, sheep were envisioned as gentle creatures, and kings tried to associate themselves as shepherds of their flocks. The word "shepherd" was used to refer to a king, teacher, judge, or deity. This is part of the *pastoralism* view of animals associated with western cultures.[33] The arguments of caring for animals has been an ongoing discussion even into the twenty-first century, and biblical references continue to be used.[27] Horses missed much of the religious connection, but their importance in culture was still very prominent in other ways. By the time of Medieval Europe, the French had several different words for horse, with each signifying a different function.[8]

As farming became more widespread, *agrarianism* became the prevailing view of the human–animal relationship.[13] The underlying philosophy of agrarianism is that living close to the land brings out the best in people and that they live a simple life. Cities corrupt people. Factories were bad and were associated with heavy smoke and pollution. Animals generally lived on small farms. Life was wholesome and in tune with nature. It was probably assumed that each animal had a name.

Around 1600, the view of animals began to change. This change is reflected in artwork of the time. Prior to that date, animals were used more as symbols or as decorations for the rich who had their portraits painted. After 1600, *romanticism* resulted in animals being depicted as the primary subject of paintings, probably because they were part of Nature, which also had a higher priority then.[13] The paintings would depict animals with emotions and pictured people showing empathy for or friendship with an animal. Animals were starting to be viewed as living beings that had personalities and emotions.

In another hundred years, the idea of *industrialism* meant "progress," and "science has all the answers" became the prevailing mindset.[13] Value was placed on productivity and efficiency, and these were associated with prosperity and making human life easier. The intensification of agricultural production paralleled mass production in other industries. Animals, then, were viewed as workers in efficient production systems, and eventually they came to be viewed by some as unfeeling machines. Their health and production were the sole measures that determined welfare. Ultimately, the way animals were being treated resulted in the first law regarding the protection of animals being passed in Britain in 1822—the Act to Prevent the Cruel and Improper Treatment of Cattle.[27,31]

A converging set of outside influences began to affect animal production between 1929 and the end of World War II. Farm laborers became scarce, and food shortages were more common. At the same time, the development of practical refrigerators and rapid transportation meant that food could go to markets far from the source, and farmers had to start competing for urban markets. Animal production became increasingly automated to minimize labor needs and yet keep costs of production competitive.

After World War II, the newfound plenty gave people time to think about things besides getting food and paying for housing. As soldiers returned, they tended to stay in urban areas. The shift away from the farm affected how people thought about livestock. In 1940, 18.5% of people were involved in agriculture.[38] By 1990, the US Census indicated just 1.6% of the population lived on the farm.[37] Today, the vast majority of Americans are at least three generations off the farm. They have little-to-no contact with farms and few good sources from which to learn about agriculture. Rightly or wrongly, perceptions about livestock care are typically based on their views of how a pet dog or cat should be treated, without realizing what is normal and/or appropriate for livestock.[9]

By 1960, there was a growing concern about the treatment of production animals. Ruth Harrison's book, *Animal Machines*, gave voice to many of the concerns. Ultimately, the power of the issues was strong enough that a special commission was put together in the United Kingdom and was charged to investigate the welfare of animals kept under intensive livestock husbandry systems—the Brambell Committee. Included in their final report (1965) was a statement that animals should have the freedom to "stand up, lie down, turn around, groom themselves and stretch their limbs."[11] This statement eventually led to the codifying of the Five Freedoms (freedom from hunger or thirst; discomfort; pain, injury, or disease; fear and distress; and to express normal behavior) in 1979. The Brambell Report also contained the suggestion that science should be used to research animal stress, pain, discomfort, cognitive abilities, and the capacity to learn a fear of humans.[13,21] Ultimately the Brambell Committee report led to the discussion of ways to evaluate animal perception and feelings. As scientists study these issues, research results are providing a better understanding of animal behavior and leading to the development of more humane handling techniques for livestock.

REFERENCES

1. Amaral AJ, Ferretti L, Megens H-J, Crooijmans RPMA, Nie H, Ramos-Onsins SE, et al. Genome-wide footprints of pig domestication and selection revealed through massive parallel sequencing of pooled DNA. *PLoS One* April 2011;**6**(4):1–12. e14782.
2. Beaver BV. *The veterinarian's encyclopedia of animal behavior.* Ames (IA): Iowa State University Press; 1994. p. 307.
3. Bollongino R, Burger J, Powell A, Mashkour M, Vigne J-D, Thomas MG. Modern taurine cattle descended from small number of near-eastern founders. *Mol Biol Evol* 2012;**29**(9):2101–4.
4. Bruford MW, Bradley DG, Luikart G. DNA markers reveal the complexity of livestock domestication. *Nat Rev Genet* Nov. 2003;**4**:900–10.
5. Clutton-Brock J. *A natural history of domesticated mammals.* 2nd ed. Cambridge (UK): Cambridge University Press; 1999. p. 238.
6. Clutton-Brock J. How domestic animals have shaped the development of human societies. In: Kalof L, editor. *A cultural history of animals in antiquity*, vol. 1. New York: Berg; 2011. p. 71–96.
7. Clutton-Brock J. The unnatural world: behavioural aspects of humans and animals in the process of domestication. In: Manning A, Serpell J, editors. *Animals and human society: changing perspectives.* New York: Routledge; 2011. p. 23–35.
8. Cohen E. Animals in medieval perceptions: the image of the ubiquitous other. In: Manning A, Serpell J, editors. *Animals and human society: changing perspectives.* New York: Routledge; 2011. p. 59–80.
9. Digard J-P. Relationships between humans and domesticated animals. *Interdiscip Sci Rev* September 1994;**19**(3):231–6.
10. Driscoll CA, Macdonald DW, O'Brien SJO. From wild animals to domestic pets, an evolutionary view of domestication. *Proc Natl Acad Sci* June 16, 2009;**106**(Suppl. 1):9971–8.
11. Farm Animal Welfare Committee: Five freedoms. http://www.defra.gov.uk/fawc/about/five-freedoms/; downloaded June 5, 2013.
12. Fernández H, Hughes S, Vigne J-D, Helmer D, Hodgins G, Miquel C, et al. Divergent mtDNA lineages of goats in an Early Neolithic site, far from the initial domestication areas. *Proc Natl Acad Sci* October 17, 2006;**103**(42):15375–9.

13. Fraser D. *Understanding animal welfare: the science in its cultural context.* Ames (IA): Wiley-Blackwell; 2008. p. 324.

14. Gifford-Gonzalez D, Hanotte O. Domesticating animals in Africa: implications of genetic and archaeological findings. *J World Prehist* 2011;**24**:1–23.

15. Giuffra E, Kijas JMH, Amarger V, Carlborg Ö, Jeon J-T, Andersson L. The origin of the domestic pig: Independent domestication and subsequent introgression. *Genetics* April 2000;**154**:1785–91.

16. Götherström A, Anderung C, Hellborg L, Elburg R, Smith C, Bradley DG, et al. Cattle domestication in the Near East was followed by hybridization with aurochs bulls in Europe. *Proc R Soc B* 2005;**272**:2345–50.

17. Groenen MA. Analyses of pig genomes provide insight into procine demography and evolution. *Nature* November 15, 2012;**491**(7424):393–8.

18. Hansen T, Forster P, Levine MA, Oelke H, Hurles M, Renfrew C, et al. Mitochondrial DNA and the origins of the domestic horse. *Proc Natl Acad Sci* August 6, 2002;**99**(16):10905–10.

19. Hiendleder S, Lewalski H, Janke A. Complete mitochondrial genomes of *Bos taurus* and *Bos indicus* provide new insights into intra-species variation, taxonomy and domestication. *Cytogenet Genome Res* 2008;**120**:150–6.

20. Hiendleder S, Mainz L, Plante Y, Lewalski H. Analysis of mitrochondrial DNA indicates that domestic sheep are derived from two different ancestral maternal sources: no evidence for contributions from urial and argali sheep. *J Hered* 1998;**89**(2):113–20.

21. Keeling LJ, Rushen J, Duncan IJH. Understanding animal welfare. In: Appleby MC, Mench JA, Olsson IAS, Hughes BO, editors. *Animal welfare.* 2nd ed. Cambridge (MA): CAB International; 2011. p. 13–26.

22. Larson G, Albarella U, Dobney K, Rowley-Conwy P, Schibler J, Tresset A, et al. Ancient DNA, pig domestication, and the spread of the Neolithic into Europe. *Proc Natl Acad Sci* September 25, 2007;**104**(39):15276–81.

23. Larson G, Dobney K, Albarella U, Fang M, Matisoo-Smith E, Robins J, et al. Worldwide phylogeography of wild boar reveals multiple centers of pig domestication. *Science* March 11, 2005;**307**(5715):1618–21.

24. Larson G, Liu R, Zhao X, Yuan J, Fuller D, Barton L, et al. Patterns of East Asian pig domestication, migration, and turnover revealed by modern and ancient DNA. *Proc Natl Acad Sci* April 27, 2010;**107**(17):7686–91.

25. Ludwig A, Fruvost M, Reissmann M, Benecke N, Brockmann GA, Castaños P, et al. Coat color variation at the beginning of horse domestication. *Science* 2009;**324**:485.

26. MacHugh DE, Shriver MD, Loftus RT, Cunningham P, Bradley DG. Microsatellite DNA variation and the evolution, domestication and phylogeography of taurine and zebu cattle (*Bos taurus* and *Bos indicus*). *Genetics* July 1997;**146**:1071–86.

27. Maehl A-H. Cruelty and kindness to the 'brute creation': stability and change in the ethics of the man-animal relationship, 1600-1850. In: Manning A, Serpell J, editors. *Animals and human society: changing perspectives.* New York: Routledge; 2011. p. 81–105.

28. McMiken DF. Ancient origins of horsemanship. *Equine Vet J* 1990;**22**(2):73–8.

29. McTavish EJ, Decker JE, Schnabel RD, Taylor JF, Hillis DM. New World cattle show ancestry from multiple independent domestication events. *Proc Natl Acad Sci* April 9, 2013;**110**(15):E1398–406.

30. Naderi S, Rezaei H-R, Pompanon F, Blum MGB, Negrini R, Naghash H-R, et al. The goat domestication process inferred from large-scale mitochondrial DNA analysis of wild and domestic individuals. *Proc Natl Acad Sci* November 19, 2009;**105**(46):17659–64.

31. Ritvo H. Animals in nineteenth-century Britain: complicated attitudes and competing categories. In: Manning A, Serpell J, editors. *Animals and human society: changing perspectives.* New York: Routledge; 2011. p. 106–26.

32. Rubin C-J, Megens H-J, Barrio AM, Maqbool K, Sayyab S, Schwochow D, et al. Strong signatures of selection in the domestic pig genome. *Proc Natl Acad Sci* November 27, 2012;**109**(48):19529–36.

33. Schubert M, Jónsson H, Chang D, Sarkissian CD, Ermini L, et al. Prehistoric genomes reveal the genetic foundation and cost of horse domestication. *Proc Natl Acad Sci* December 30, 2014;**111**(52):E5661–9.

34. Schwabe WW. Animals in the ancient world. In: Manning A, Serpell J, editors. *Animals and human society: changing perspectives.* New York: Routledge; 2011. p. 36–58.

35. Serpell J, Paul E. Pets and the development of positive attitudes to animals. In: Manning A, Serpell J, editors. *Animals and human society: changing perspectives.* New York: Routledge; 2011. p. 127–44.

36. Szűcs E, Geers R, Jezierski T, Sossidou EN, Broom DM. Animal welfare in different human cultures, traditions and religious faiths. *Asian-Australas J Anim Sci* November 2012;**25**(11): 1499–506.

37. United States Census Bureau: Farm population, farm earnings, and agriculture. http://www.census.gov/prod/2002pubs/00ccdb/cc00_tabA10.pdf; downloaded May 31, 2013.

38. United States Census Bureau: 1940–2010 How has America changed? http://www.census.gov/1940census/pdf/infographic1_text_version.pdf; downloaded May 31, 2013.

39. Vilá C, Leonard JA, Götherström A, Marklund S, Sandberg K, Lidén K, et al. Widespread origins of domestic horse lineages. *Science* January 19, 2001;**291**:474–7.

40. Warmuth V, Eriksson A, Bower MA, Barker G, Barrett E, Hanks BK, et al. Reconstructing the origin and spread of horse domestication in the Eurasian steppe. *Proc Natl Acad Sci* May 22, 2012;**109**(21):8202–6.

41. Wu G-S, Yao Y-G, Qu K-X, Ding Z-L, Li H, Palanichamy MG, et al. Population phylogenomic analysis of mitochondrial DNA in wild boars and domestic pigs revealed multiple domestication events in East Asia. *Genome Biol* 2007;**8**(11):R245.1–12.

42. Zeder MA, Hesse B. The initial domestication of goats (*Capra hircus*) in the Zagros mountains 10,000 years ago. *Science* March 24, 2000;**287**:2254–7.

43. Zohary D, Tchernov E, Horwitz LK. The role of unconscious selection in the domestication of sheep and goats. *J Zool* June 1998;**245**:129–35. London.

Chapter 2

Behavior as it Relates to Handling

Never approach a bull from the front, a horse from the rear, or a fool from any direction.

Author unknown

I tell Sasa what to do, and Sasa tries to understand. His neck curves away from me, as arched as a chess knight's, and at the end, those two eloquent white ears. Like radar dishes, they swivel to concentrate on me, or on another horse nearby, or they drop sideways in semaphore for – what? – concentration or confusion?[33]

Success in working with livestock is not related to the intelligence or lack of it on the part of the animals. It really depends on the animal's ability to try behaviors until something works.[62] The process of getting an animal from point A to point B can be a distressful, reactive, trial and error attempt in which the human finally lucks out at getting the job done, or it can be a stress-free experience for the human and the animals. Which type of experience it is will depend on the human's understanding of how an animal perceives its world, how animals learn, the relative importance of its social peers, and what motivates it to move. With that knowledge, appropriate techniques can be used to humanely create an outcome that is both effective and efficient.

THE SENSES

Domestic livestock species came from wild species that were prey for other animals. Each had to develop senses that could detect the approach of a predator and skills to avoid predation. As a result of this need, the senses of the livestock species are similar to each other but quite different than those of humans. In day-to-day life, the senses are used in combinations rather than individually. As an example, horses recognize approaching individuals by using a combination of senses. Hearing might be the first to call attention to where a foal or mare is, but vision becomes important for approaching and recognizing whether it was the right mare or foal. Finally, olfaction and touch result in reuniting the bond pair.[59,154]

Efficient Livestock Handling. http://dx.doi.org/10.1016/B978-0-12-418670-5.00002-0

13

Irrespective of which of the livestock species is being discussed, what handler action is intended, or what behavior is occurring, animals depend on their senses for environmental input. The senses serve as the primary filters for incoming information. An animal must be aware that there is something if it is to react to or learn from the experience. As an example, a blind horse would not see a bird suddenly fly up and so would not shy because of the rapid movement. In this case, the lack of sensory perception means the stimulus does not get to the brain, and so the horse cannot use its memory or instinct to react to the stimulus. Different animals have different sensory capabilities and different emphasis based on those capabilities. The sensitivity of a dog's nose compared to that of a human illustrates this point. When one sense is diminished, the other senses tend to be heightened. The blind horse alerts more easily to sounds and smells to compensate for not being able to see.

Once this incoming information passes through a sensory filter and reaches the brain, it is processed to determine if a reaction is necessary and what that action might be. As a secondary filter, the brain is subject to variability. First, it has to be aware that stimuli are incoming, something that can be disturbed with brain injuries. Then it has to interpret the stimuli, and interpretation can be influenced by what humans call "moods." These are influenced by the time of day, previous experience, specific environments, and weather conditions. As an example, on windy days the sensory system receives a lot of input due to blowing scents from all over. This puts a horse on high alert so other messages coming to the brain are more likely to be summed as a potential danger rather than be lightly regarded, as would happen on a calm day.

Vision

Humans use vision as the primary sense. For livestock species, vision, hearing, and smell are complimentary and about equal in importance. Horses, cattle, sheep, and goats have visual systems that are about the same as each other, but different from humans in several aspects. Livestock species use the differences in brightness, motion, distance, texture, orientation, and to a small degree color to survey their environments. The majority of studies have been conducted on horses, so most of the data that follows uses them as the primary model. Pigs are different and will be discussed separately.

Visual systems of livestock are designed for scanning the environment in a panoramic fashion in daylight and dark rather than for defining sharp details. Detecting a predator that is sneaking up in time to allow escape or prepare to fight is the important thing. The most evident variation between predator and prey is the visual field (Figure 2.1). Each eye of prey species can see an arc of approximately 200°–210° around the body at one time. There is an area in front of the animal where these monocular fields overlap to give the animal binocular vision.[97,110] The binocular field is typically between 65° and 80° going straight forward, perhaps somewhat less in sheep.[109] This is where the most accurate depth perception occurs because of the overlapping, stereoscopic views. While there is some depth perception using monocular vision, it is about one-fifth the strength of binocular

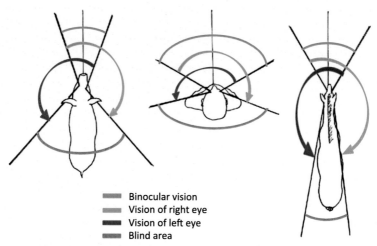

Binocular vision
Vision of right eye
Vision of left eye
Blind area

FIGURE 2.1   A comparison of visual fields of swine, humans, and horses.

vision.[39] When approached, a horse might suddenly appear more alert merely because it is changing from monocular to binocular vision. The shape of the head and position of the eyes are the major determinants as to what the visual field will be for each individual. Animals with eyes set on the side of the head, as in a horse, have a larger total field than those having eyes set in the front, as for humans. In livestock species, fine-tuned depth perception is less important than it is in people. Of the animal species, horses might need the best depth perception because of their mobility and ability to jump. Even for the horse, their depth perception is significantly less than humans. They can see a difference in height of about 4.5 inches (9 cm) from a distance of 6.5 feet (2 m), while a human could detect a difference of about one-eighth of an inch (a few millimeters).[141] The remaining area directly behind the animal's head and body is called the blind area, but that approximately 20° field can easily be scanned with slight side-to-side movements of the head and by the position of the head relative to parts of the body that might block the eyes (Figure 2.2). It is only in this blind area that objects are not easily recognized.[48] There are also blind areas located for a few feet immediately in front of the forehead, directly under the head, and above and below the body. It is estimated that the vertical plane of vision from each eye is about 178°, creating an almost complete sphere of vision.[97] Limiting the field of vision can be stressful. Horses that wear blinkers will show an increased heart rate when there is an unfamiliar sound but will have a decreased rate in the presence of a familiar visual stimulus.[27]

The lens of the eye adjusts to give the sharpest vision as distance between the animal and the approaching image changes. It becomes thicker or thinner. At approximately 1.5 feet (0.5 m), the lens no longer can sharpen the focus of something approaching, so the horse may back away or move its head to try to see better.[39]

The pupil shape of domestic livestock species is a horizontal rectangle (Figure 2.3). This gives the animals a much wider area of visual perception for

FIGURE 2.2   The lowered head allows this horse to use its monocular vision to scan the lateral horizon for potential threats, such as a rider trying to catch it, and use binocular vision to assess suitable food and safe footing.

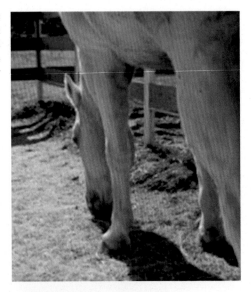

FIGURE 2.3   Goats have an elongated, rectangular pupil, as do horses, cattle, and sheep. *Photo compliments of Bridget and Gary Sebesta; used with permission.*

View as seen by humans

View approximating that as seen by horses, cattle, sheep, and goats

FIGURE 2.4   Concept comparison of the width of visual perception of humans and domestic livestock. *Based on images by Nickolay Lamm about cat vision in consultation with Drs Kerry Ketring, D.J. Haeussler and the Veterinary Ophthalmology Group at the University of Pennsylvania; used with permission.*

scanning than they would have with round pupils (Figure 2.4).[45,109] Light that enters the eye through the pupil travels to the back to stimulate various cells of the retina. These photoreceptor cells at the back of each eye vary in concentration depending somewhat on location.[30,97,140] Near the edges of the retina in horses, there can be as few as 16–304 cells/mm², but there is a band near the optic disc where these nerve cells are particularly concentrated called the "visual streak." This area parallels the shape of the pupil and can have greater than 6500 cells/mm².[30,97,140] Images that land on the visual streak are those experienced with the sharpest vision and so are associated with binocular vision.[45] This is best accomplished with the images entering near the bottom of the eye, just like the lower part of bifocal glasses. It explains why a running horse carries his head raised with nose extended somewhat to see distant objects and why one walking over a set of logs would bend the head and neck downward.[51] Within reason, muscles that move the eye compensate for the head position so that changes do not affect the horizontal position of the pupil.[6] While true in horses, it has been suggested that it may not be true in cattle.[115] Rotation of the eye to maintain the horizontal orientation of the pupil will allow vision to remain normal whether the animal is grazing or browsing. Extreme head positions, such as the Rollkur posture of hyperflexion, have been shown to be stressful,[146] affecting vision and possibly the ability to breathe. Horses show behavioral signs of discomfort in these positions characterized by tail swishing, fighting the bit, and attempting to buck.

The photoreceptor cells of the eye consist of two kinds—rods and cones. In humans, there are approximately nine rods for every cone.[45,47] In the livestock species, the ratio is approximately 20:1.[45,98] This hints at two differences from humans: less use of color vision in animals, because cones are associated with that, and better vision at night, since rods are the cells adapted for low light. Humans are particularly dependent on the cones of the eye for color vision. People are considered to be trichromats because the cone sensitivity peaks at three different light wavelengths. The red peak, as detected by long wavelength-sensitive (L) cones, is about 560–565 nm. Medium wavelength-sensitive (M) cones (530–535 nm) cover the yellow spectrum area, and the short wavelength-sensitive (S) cones (430–440 nm) respond to the bluish colors.[86] Horses, cattle, sheep, and goats are dichromats, with S cones that detect the light of wavelengths from approximately 439–456 nm and middle-to-long wavelength-sensitive (M/L) cones for about 537–557 nm wavelengths.[15,37,49,87,97,103,118,120,141] The result is that, like most mammals, livestock species basically see a continuous scale of colors ranging from blue through yellow (Figure 2.5).[120] This would be similar to a human with a red-green color blindness. Color perception can also be affected by the amount of illumination of the object being tested. In the past, studies to determine if animals could see colors suggested they could, but they failed to take into account how the shine off a red object compared to the drabness of a dark green might be the real signal the animal was using as a cue. It has been shown that while horses do perceive colors across large luminance differences, the results still indicate dichromatic vision.[86,104] Instead

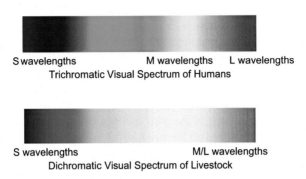

FIGURE 2.5   Color perceptions of humans and of domestic livestock.

of detecting things by color, the relatively high number of rods suggests that animals that are active at night are better at using contrasts of texture and the plain backgrounds for safe movement and predator detection.[47] Livestock tend to avoid sharp contrasts in pathways because of this, which explains how cattle regard cattle guards. Unless there is time to investigate the cause of that contrast, livestock will try to avoid having to go over or past such changes.

In their natural environments, livestock species eat day and night, with concentrated periods typically associated with dawn and dusk. The most important aspect of vision is to warn the individual of approaching predators even in low light conditions and to assess ground conditions for both eating and escape. The eyes of livestock are not particularly adept at making rapid transitions between bright and dark locations. It has been reported that accommodation in horses going between extreme brightness to darkness can take over 30 min to have complete adaptation.[47,141] This explains their reluctance to move between areas where illumination differences are great, such as loading into a dark trailer or going from outdoors into a dark, unfamiliar building. In nature, darkness comes slowly, and color vision gives way to black and white in livestock at about the same visual light threshold as happens in people because the sensitivity of the cones is similar.[121] Although rods are very important in daytime vision for detecting movement, they become the only photoreceptors active in low light conditions.[125]

At the back of the eye is a reflective layer, the tapetum lucidum (specifically the choroidal tapetum fibrosum), in horses, cattle, sheep, and goats. Deer eyes glow green at night when the tapetum lucidum reflects car headlights. The reflection from human eyes is red because we lack this tapetal layer, as do some individual animals. The tapetum enhances vision at night but complicates it somewhat during the day.[97,99,125] When light enters the eye, it will trigger a photoreceptor on the retina and then will be reflected from the tapetum so that it can trigger additional receptors. In essence, the tapetum serves as a mirror and the amount of light is magnified. This results in the animal's ability to see at lower light levels than humans and other animals which do not have a tapetum can, but it also decreases the resolution of incoming images, blurring them. As a result, viewed objects, particularly smaller ones, are harder to identify, and moving ones are more likely to trigger the fight or flight reactions typical of each species. During daylight, the pupil will need to constrict significantly to minimize the amount of light that enters. In general, though, the images will be blurred, and the ability to see detail is less than that for humans (Figure 2.4).[15,140]

Vertical objects are more easily detected than horizontal ones.[115] This may occur for several reasons. It might make predators more easily distinguishable from the horizontal background or heights more easily judged compared to distances because of limited binocular vision. Specifics about the visual acuity of animals in general and horses specifically is controversial because of conflicting data.[97] Tests for visual acuity typically use vertical black and white strips of varying widths. Test results suggest that in horses, the acuity is approximately 20/30, which is a little better than the human standard of 20/20.[46] That means they are farsighted, seeing at 30 feet what humans can see at 20 feet.

The visual sense of pigs differs from that of other livestock species and in some ways approximates that of humans. The visual field is more typical of an omnivore than of a prey species. Their forwardly placed eyes result in a relatively large blind area behind their head, approximating 50° (Figure 2.1). The binocular area is only about 30°–50° and is limited by the shape of the snout.[109] This results in a monocular field of about 115°. Pigs have a round pupil and they lack a tapetum lucidum.[99] These last two adaptations would result in pigs being active when there is enough light for humans to also see. Like other livestock species, however, pigs are dichromats.

Vision has been shown to be important in a number of behaviors, such as the initiation of a bull's sexual response.[35] Studies in pigs and cattle showed that the recognition of familiar handlers was completed primarily by the color or contrast of color (light/dark) of the coveralls being worn.[72,122] This emphasizes the importance of consistency when working with livestock species.

A different eye, and thus a different part of the brain, is used when distinguishing familiar from potentially harmful objects. In cattle, novel stimuli are preferentially viewed with the left eye, while the right monocular field becomes favored with familiarity.[119] In horses, the preferential eye for novel stimuli is also the left one, particularly in flightier animals.[82,124] This is supported by the findings in other social species, including humans. The proposed explanation is that the right hemisphere of the brain, with input mainly from the left side of the body, developed the role of integrating ongoing information to immediately detect and respond to novel and unexpected events.[25,119] It is interesting to speculate whether mounting horses from the left side happened by chance or was the result of learning that it was safer or easier. Calmer-natured horses do not show this left viewing preference and will use their right eyes even when exploring novel objects.[23,82] They do, however, have a preference to use their left eyes for interactions with humans, even when they have been specifically trained for human interaction on both sides.[31]

Hearing

The second sense humans tend to think is most important is hearing. For those species where audiograms have been done, the resulting curve is similar in shape to that of humans and shifted slightly to the right (Figure 2.6).[54,56] Livestock generally do not hear the lowest sounds we do, but they can hear slightly higher ones (Table 2.1). The audible range for young adult humans is approximately 20 Hz to 20 kHz, with a peak sensitivity of 1–3 kHz.[141] For a horse, the range is from 55 Hz to 33.5 kHz, with the best sensitivity between 1 and 16 kHz, which is within the normal range for most equine vocalizations.[56,156] As can happen in people, older horses may experience a significant reduction in hearing ability.[153] Cattle hearing ranges from 23 Hz to 35 kHz, with best sensitivity at 8 kHz.[56] As with humans, the unique characteristics of the cattle

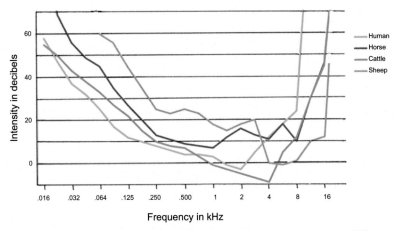

FIGURE 2.6   Comparative audiograms of humans, horses, cattle, and sheep.[54,56]

TABLE 2.1 Range of Hearing for Humans and Livestock[13]

Species	Lowest Frequency Detected (KHz)	Greatest Sensitivity (KHz)	Highest Frequency Detected (KHz)
Humans	0.031	8	17
Horses	0.055	2	33
Cattle	0.024	8	40
Pigs	0.040	8	40
Sheep	0.125	10	40
Goats	0.070	2	40

vocalizations are useful for identifying individuals.[155] Exposure to the calls of cows will increase bull libido as well.[35] Pigs have a hearing range that has been reported to be from 42 Hz to 40.5 kHz, another species that can hear outside of the human range.[38]

The mobility of each ear in livestock species suggests that hearing is an important sense, but studies have shown that the use of the ears to localize the source of the sound is inversely proportional to the width of the field of best vision.[59,141] Large monocular visual fields mean less use of hearing for localization. In prey species except the pig, the hearing probably is used to draw attention to something, but other senses take over to determine the actual source.[60,154] It has been suggested that interaural distance may play a role in sound localization because of sound arriving at different times to the two ears. In humans, and possibly pigs at low tone frequencies, the different arrival times at each ear will aid in localizing sound sources.[58] In other livestock species, which actually have a large interaural distance, sound localization is poor.[55,57,60]

Exposure to loud noises will increase the heart rate of individuals in a number of species to a level consistent with having been physically hit. Studies have shown that the strongest and most consistently aversive response to any sensory stimulus is to sounds.[65] Cattle have the most acute hearing of the livestock species and can hear softer sounds than humans can hear: something of high significance when humans work with them. Loudness is exaggerated. Holstein cattle are more sensitive to sound than are beef cattle, and individuals that were sensitive to sudden, intermittent sounds were more likely to become agitated during handling.[80]

Olfaction

The sense of smell in animals is not well understood by humans because our own olfactory system is so rudimentary. This makes setting up and interpreting controlled studies very difficult. Humans are considered to be microsomatic, having very small olfactory bulbs in the brain. The livestock species are considered to be macrosomatic because of the relatively large size of the olfactory bulbs, including numerous folds which increase the surface area even more. The density of olfactory receptor cells remains constant over the surface area of the olfactory bulbs, so the larger the surface area, the larger the number of receptors.[125] We can appreciate that the sense of smell is very important to livestock species because it is used in many behavioral contexts. Domesticated horses vary in their use of olfaction from their wild relatives. Wild Przewalski stallions will direct sniffing toward the genital regions of others they are investigating, while modern horses are significantly more likely to sniff the noses or bodies of strangers.[18] Evidence for the differentiation between urine or body odors of individuals is less clear.[63,130] The volatile components of horse urine have chemically detectable differences based on the stage of the estrous cycle and the sex of the horse, so it is possible that horses can detect these differences.[63]

Horses are also able to distinguish the feces of competitors from nonthreatening herd mates through smell.[73] If horses are able to distinguish competitors, then it is likely that they can differentiate between individuals. Pigs and sheep have been shown to recognize individual conspecifics; however, goats and calves have not, at least yet.[3,4,91]

Body odors may also be significant to livestock species. Stallions and geldings generally sniff an area prior to rolling and they tend to roll in the same area previously used by another horse.[123] Bulls show an increase in smelling and tasting behaviors toward proestrus cows beginning about four days before estrus, suggesting one or both of those behaviors can detect the subtle changes which occur.[35] Pigs have a particularly keen sense of smell, as is shown in those used to find wild truffles.

As with other senses, olfaction plays a role in identifying the approach of predators for prey species. The presence of the odor of a predator increases vigilance without fright. It will cause an increase in the heart rate, however, if another avoidance-inducing stimulus (such as a sharp noise) is also present.[17] It has been suggested that for horses, and potentially for cattle, stereolfaction may help with the location of an odor's source due to the widely spaced nostrils.[125]

Flehmen, the elevation of the upper lip, is seen in many species and is associated with a second olfactory system (Figure 2.7). The behavior is expressed by the animal extending its neck, raising its nose, opening its mouth slightly, and lifting the upper lip. Flehmen is shown by sheep and goats too, but the attachment of the upper lip in its center makes the lip elevation less obvious. The posture of the head opens paired ducts in the mouth just behind the dental pad or incisor teeth, ones in the nasal passages, or both. This allows the passage of nonvolatile materials into them.[22,76] Species that have the openings in the mouth typically lick the area they are investigating before showing the flehmen. The oral/nasal ducts lead to a pair of cigar-shaped sacs, called the vomeronasal organs, which are located on the floor of the nose. Each of these is lined with an olfactory epithelium that responds to specific types of odor molecules associated with reproductive behaviors. Nerves leaving the vomeronasal organs go to the accessory olfactory areas of the brain to trigger innate responses rather than the responses learned in association with the primary olfactory bulbs.[21] Of the livestock species, horses are the only ones that lack ductal openings in the mouth (they only have the nasal ones).[21,29,123,131] As a result, they seldom show the licking behaviors. Horses are also more likely to follow flehmen with urinary marking than with other courtship behaviors, suggesting that the flehmen response is associated with sensory priming for reproduction.[68,125,131] Bulls housed in a bull stud react strongly to the urine of other bulls with sniffing, licking, and flehmen, but only when not in an area where they are collected.[107] The likelihood that flehmen is shown more often in association with anestrous urine and postcopulation is evidence that the behavior is for

FIGURE 2.7 Flehmen behavior allows volatile molecules associated with sexual and social behaviors to enter the secondary olfactory system through the vomeronasal organ. For this horse, it was useful to examine a cracked tooth. For the bull, it was used to check the estrous status of a cow. *Bull photo compliments of Allen Russell; used with permission.* © *Don Höglund (2015).*

priming and maintenance of sexual interest functions.[68,77,107] This is apparently true in other species as well because castration results in a reduction in the frequency of flehmen.[52]

Touch

We do not know much about the sense of touch in our various livestock species other than the general pattern of nerve distribution. Which nerves innervate the various muscles and skin areas is generally well studied, particularly in horses and cattle. The general types of nerve endings suggest that animals feel pain, pressure, warmth, and cold. This differs little across species. The intricacies of touch are not well studied, although horses certainly react to a fly landing on its body hair, and most animals are particularly sensitive around their lips. The technique used,

location and amount of area touched, timing, amount of warning, and angle manner of the approach are critical elements in how the animal will respond (Figure 2.8).[88] One unpublished study indicated that the horse's sensitivity of touch in the areas where a rider's legs would be is greater than on a human fingertip.[81,125] Pigs tend to respond to tactile stimuli in a positive manner, particularly if it comes from a person known to the pig.[65] Similar results were shown in finishing cattle. When a gentle touch was applied during their last five weeks, they showed less human avoidance and slightly less distress at slaughter.[111] Meat quality, however, was improved more significantly when the gentle handling occurred early in life instead.

The use of a twitch in horses is a common practice with interesting associated research. Studies have shown that the heart rate does not increase as would be expected with a painful procedure.[78] The horse will typically drop its head, relax its eyes, and appear somewhat sedated. It has been suggested that these responses might be associated with an acupuncture-like reaction.

Pheromones

Pheromones are substances that are produced by one animal that carry information to another animal. By definition, then, a pheromone could be something detected by smell, including by the vomeronasal organ or by taste. Saliva and other sources of pheromones function in livestock reproduction, nursing, and social interactions in cattle, pigs, sheep, and goats.[13,117] Wild ruminants have active glands between the digits of their feet, but domestic ruminants only have remnants of those glands that may or may not be functional.[13] In sheep, pheromones in the wool, wax, and urine of a ram can stimulate ewes to ovulate. Buck goat pheromones accentuate doe estrus behaviors. Boar odor improves ovulation and sperm transport. The specific role of pheromones in cattle is a little less clear, but it is believed to speed up the onset of puberty in heifers and enhance early resumption of ovarian activity in postpartum cows.[117] The nose-to-nose contacts of these species are more of a nose-to-mouth contact, allowing the transfer of chemical information.

SOCIAL BEHAVIOR

As illustrated in the true story at the beginning of this chapter, knowledge of the social structure and group dynamics of a species is important to understanding how to best work with those animals. Domestic livestock species are social animals, meaning that they live in groups rather than alone. Working with them, then, should take into account that individuals will want try to stay with the group and that group members will tend to move as a unit. Cutting horses demonstrates the difficulty of separating an individual cow from its herd.

Group living has advantages and disadvantages. For animals that evolved as a prey species, there are survival benefits to having other potential meals around you and your offspring. This, in turn, means less time is needed for surveillance activities and more time can be used for grazing. On the other hand, there can be significant stimuli from higher ranking individuals, making close proximity stressful, and too many individuals can result in food competition. In the wild, individual animals would be able to enter or leave herds/flocks with relative ease. They can allocate space both within and between groups.[83] Managed herds are generally less affected by predators, but stocking rates and access to high value items like food can complicate the dynamics of the herd or flock as a social unit.

Social Group Dynamics

All the livestock species being discussed live in groups, and cohesion of group members is tested when individuals are separated out. In general, isolation is a strong stressor for livestock, and it tends to result in increased agitation and excitement. Visual contact is often sufficient to negate the stress response. As an example, horses taken for a trail ride become very distressed, especially the first time they leave the barn alone. For that reason, young horses are typically accompanied by an older, experienced horse until the younger one becomes familiar with the places it will go and the types of things it will see. Over time, many horses learn to go alone, particularly if they are trusting of their rider. Cattle, too, adapt somewhat to being alone. If separated from herd mates, cattle typically show increased vocalization, vigilance, and walking.[41,94] What will change over time is the degree of stress, as indicated by cortisol levels. Initially, these levels are elevated. With repeated separations, the behaviors do not change appreciably, but the cortisol levels approach normal.[94] Calves raised alone show less distress later when separated from their peers.[12] On the other hand, there is a calming effect with the presence of larger groups on individuals undergoing stressful situations.[137]

Members of a herd will establish social orders within that group. We typically talk about "peck" or "dominance" orders, but *social order* is actually a better term to describe the relationships between individual animals. "Dominance" implies a ranking with a "dictator" at the top. In reality, there is a relative ranking between individuals that is quite stable and subtle. Interactions are not just determined by rank but can depend on other things in the environment, the species involved, and the value of the resource. As an example, one would expect that social rank and the relative position of other horses will determine that the higher ranking one would get access to high value food first.[75] That does not always happen. A high ranking horse that can displace a pasture mate when the two are about to enter a barn may defer to that same pasture mate when it comes to accessing a pile of hay. In getting access to food, higher ranking cattle displace lower ranking ones as often as they are displaced by the lower ranking ones.[132]

Social orders are important for group harmony and stability. They result in the minimization of psychological and physical stress in competitive situations.[108] Threats replace serious aggression in disputes. Deference, not aggression, is used as the primary indicator of rank. Young animals generally come in at the bottom of the social order, showing deference to older individuals. Over time, the position within the herd/flock will rise as higher ranking individuals are sold off or die. Animals that are added into an established group will eventually find a location within it. Regardless of what methods are used to introduce unfamiliar horses, the frequency of aggression is the same, with resident horses being the most aggressive.[53] The factors that determine the position within a flock or herd are primarily dependent on height and weight, with bigger being higher ranking. The presence of horns, age, and sex of the introduced animal can also be determining factors. Once established, the social rankings remain relatively stable over time, and even changes in appearances, as would happen with late dehorning, usually do not make a difference.[42,135,149]

Social rankings play a role in herd and flock management. Overcrowding in pens will put psychological stress on lower ranking individuals because they encounter higher ranking ones more often.[92] As a result, the growth, reproduction, and health of lower ranking individuals can be compromised, and aggression will increase. When animals come to or leave groups, social orders can be disturbed. The reintroduction of an animal that has been away from the group for a period of time can be treated in different ways depending on the length of time gone and the species involved. In some cases, the animal is treated as a stranger, while in other cases it comes back into its original place within the social order.

In horses, the normal herd is a feral group headed by a stallion and consisting of a few mares and their foals, yearlings, and two-year-old offspring. The stallion's rank is context dependent, and he may not always have the highest ranking because of the time he spends guarding the edges of his group.[88] The young stallions are usually driven off by the time they are three years old, and fillies leave when they first come into estrus. The group dynamics are in a constant state of flux with young animals leaving and new ones showing up, at least on the periphery.[81] This arrangement has been artificially distorted when horses are under human management. New individuals are constantly being added or removed from boarding facilities. Stallions are isolated from mares except on a few breeding farms. They are often stalled individually with the only contact being between bars on adjoining stalls, at most. If later put in pasture with other stallions only, the stalled horses show more aggressive behavior that do stallions that were raised in a group with other stallions.[16] When no stallion is present in a group, the social structure in controlled by adult mares.[129] Geldings are not found under natural conditions and have introduced a new element to group dynamics. Without a stallion in the group, the geldings tend to become part of a subgroup with the juveniles in the herd, while the mares form their own higher ranking subgroup.[129]

The social order within a herd of horses is linear-tending (Figure 2.9). Individuals that share a position or form a triangular relationship are more likely to be in the middle of the social order.[88] Foals tend to develop a position that is similar to that of their dam. Birth order also has an effect on a foal's rank preweaning but not postweaning.[151] Bachelor groups of feral horses also have this linear-tending pattern. Within herds, subgroups tend to form based on "preferred associates" that spend the most time near each other.[71] Foals of preferred associates also tend to preferentially associate with each other.[151]

The human–horse relationship enhances a horse's ability to cope with novel environments. Young horses familiar with people have lower heart rates and will approach more readily if they have been handled fairly intensely or if they have been housed individually rather than in a group.[66,130]

A ⟶ B ⟶ C ⟶ D

Linear

B
↑
A ⟶ ↕ ⟶ D ⟶ E
↓
C

Linear-tending with a shared position

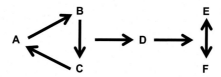

Complex

FIGURE 2.9   Examples of social orders seen in animals.

Cattle have the most complex social structure of domestic animals. Social hierarchies have been demonstrated in beef steers as young as 250 days of age and perhaps earlier in dairy calves.[134] They can have linear, linear-tending, triangular, or other diverse, complex relationships (Figure 2.9).[8,42] In addition, there are subgroups within the herd. For example, individual cows are usually part of multiple subgroups with a different relationship in each. These subgroups fit into the preference for cattle to spend time with a few favorite conspecifics. As an example, cows will spend less time resting and grazing when alone or with large groups, compared to when accompanied with two to five herd mates.[136,138] These close herd mates are the ones most likely to groom each other too. In established herds, the closest relationships tend to be based on matriarchal families, such that a mother cow and her offspring have the closest relationships.[116] Cows in estrus, heavily pregnant, or sick often disrupt the established social orders on a temporary basis.[8,36]

Groups consisting of bulls only are more likely to have a linear social structure. In these groups, mounting is more likely motivated by dominance rather than by sex.[64] This is the likely situation in buller steers too. Factors that have been shown to be associated with buller steers being mounted by pen mates include submissive behavior as well as pheromones, warm weather, group size of over 200 head per pen, and stressful events.[10] In mixed sex groups, young bulls begin to challenge the cows, such that by three years of age they are higher ranking than all the females.[34,108] Studies suggest that testosterone might result in increased dominance ranks because of the reduction of fear.[1] Social ranking and weight gain are also positively correlated.[126] Higher ranking bulls will sire a higher percentage of calves provided they are reproductively normal.[102,108]

Voluntary entrance into a milking parlor is not associated with social rank but is related to milk yield instead. Higher yield cows enter first, while lower yield cows and cows with higher somatic cell counts enter later.[102,114] There is also no correlation between dominance and milk yield.[1,24] When new cows are introduced into a dairy herd, they will be involved in twice as many altercations as resident cows during the first week, but milk yield is only reduced during the first day.[108] Previous social experience, the number of animals involved in the mixing, and the group composition influence how smoothly these introductions happen.[11] Calves are more malleable in these situations, and calves reared in groups are more socially confident than those reared individually.[11]

The social relationship of dairy cattle to humans has received a lot of study because of the significant interaction that occurs. The type of housing in which calves are raised does not affect ease or difficulty of later handling, nor does it affect milk production.[93] While it might be expected that a larger herd size would reduce the amount of positive relationships between cattle and humans, it does not have to be the case. The intensity of contact can negate that potential. Of more significance is the personality and attitudes of the people working with the cows.[148]

Sheep are the most difficult of the livestock species in which to determine social orders because they are relatively passive to challenges. Social rankings are related to age and weight, with the ram, if present, or oldest ewe being the highest ranking individual and usually the leader of movement.[64,85] Unlike other livestock species, the change of appearance, as happens with shearing, may result in a reduction in the social ranking of an individual.[64] Working sheep is made relatively easy by remembering that they are extremely difficult to separate from each other. Studies suggest that it takes at least four individuals together to have predictable behavioral responses.[85]

Goats have a stable, linear social order in which the most aggressive animals are higher ranking.[5] When food is scarce, the ranking influences what a goat eats, but that is significantly less observable when food is plentiful. That being said, meat and milk animals that are the most productive come from the middle of the social group, rather than from high or low ranking animals.[5] The most dominant animals are not necessarily the herd leaders. While the billy might be the dominant goat, the oldest nanny is usually the herd leader.[43]

The introduction of older goats into a herd can be problematic. Fighting is usually involved in settling social ranking between individuals. In some cases, the dominant-subordinate ranking is never settled, while at other times, the goats at least tolerate each other's presence.[43] Aggression is less likely to be initiated by the newly introduced goat when the confronted goat has herd mates with it. Instead, the new goat is more apt to receive aggression. This suggests that the presence of peers might be advantageous when a goat is being introduced into a group of unfamiliar goats.[101] Goats also tend to form matriarchal subgroups within the herd. Even after long separations, a returning goat will usually seek out related individuals.

Social ranking for pigs is determined within the first few days after birth and is based on which teat is preferentially nursed. The term "sucking from the hind teat" comes from pig behavior. The teats located closer to the hind legs have a poorer blood supply and thus produce less milk. Piglets that suckle farther back will grow a little slower, will have slightly less energy, and will more likely be kicked. Because size is a factor in determining social rank, the larger piglets in the front will become the higher ranking individuals within the group.[44,64]

Feral pigs are quite flexible in their social structure, living in groups varying from a few sows to groups as large as 80 animals.[44] In domestic herds when older pigs are combined into groups, aggression is often part of establishing the social rank order, and it indicates social stress. Other indicators of chronic social stress as related to stocking densities include reduced immunity, increased fever response, skin lesions, increased activity levels, and a reduction in the frequency and duration of feeding bouts.[143] Those animals ranked in the middle of the herd have the highest cortisol levels, indicating chronic stress. The ones that are low ranking have the highest levels of brain endorphins, indicating the least stress.[157]

Because mixing piglets from different sources typically involves young pigs being grown to market weight, combining litters in a location unfamiliar to them reduces the intensity of any fighting.[44] Castration not only increases feeding activity, it reduces aggression but only around three months of age.[139] Nosing of other pigs can have a social function and is not related to dominance relationships.[14]

Reactive (Social) Distances

Distances between an individual animal and another animal that is approaching will cause the one being approached to react in one of several ways based on who is coming and how far away it is. The distances between them have been called reactive, or social, distances. There are several distances and spaces that relate to domestic livestock. This is also an important subject when it comes to understanding how people can most efficiently work with these animals.

Using horses as an example, the *perceptive distance* is the distance between an animal and whatever is approaching it when the animal first becomes aware something is coming. Once it has spotted the approacher, the horse will move a step or two forward toward the approacher, and that small distance is called the *approach distance*. If the approacher continues to come toward the horse, the horse will eventually become uncomfortable and flee. The distance between the encroachment and horse when it is ready to flee is called the *flight* (escape) *distance*. In horses, this distance is 10–16 feet (3–5 m).[149] The specific distance will vary based on previous experience, what is approaching, and individual temperaments. In sheep unaccustomed to humans, the fight distance is approximately 15 feet (4.7 m), but after gentling it is reduced to 4 feet (1.3 m).[50] The flight distance is particularly important to understand for humans who work with animals, as will be discussed in greater detail later in chapters 5–9, relative to working with each species. In theory, the flight distance forms an evenly spaced area around the animal, but the limits of vision to the rear of the animal make it more likely unevenly distributed around the animal (Figure 2.10). Once the horse flees, it will withdraw a certain distance (the *withdrawal distance*) before presumably reassessing the situation, since running too far in retreat could take it from safety into another dangerous situation. While the flight distance is about the same regardless of how the approacher moves, the withdrawal distance is not. Active (as in swinging a rope or the arms) and especially rapid approaches will result in the animal moving away faster and farther.[9] For wild horses, the withdrawal distance is approximately 325 feet (100 m).[149] If, however, the horse is not aware that it is being approached or it is cornered, the approacher can continue to cross the flight distance and eventually reach the horse's *critical distance*. At this point, the horse will attack instead of trying to retreat. There is a variation between individuals and species as to the exact length of each reactive distance and how severe the attack will be. Bulls, for instance, would show a stronger reaction than would most cows.

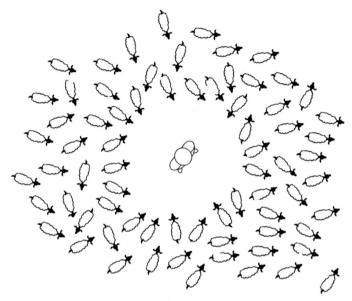

FIGURE 2.10 Sheep stay just outside the flight distance from a worker in the pen if they are not used to close interactions with the person. © *Don Höglund (2015).*

The *personal,* or *individual, distance* is the space immediately around the individual into which the animal allows what humans would call "close friends." Just as people are not comfortable having strangers walk up and stand nose to nose, animals avoid that as well. For production animals, the stocking density is directly related to personal distance. While growth is typically factored into the determination of appropriate stocking density for a given space, the hidden consequences of "psychological comfort" tend to be neglected. Relative to cow to cow interactions, lower ranking cows have a large individual distance, preferring to keep approximately 66 feet (20 m) around them clear of high ranking cows. In drylots, this means that dominant cows are more likely to intrude into another cow's space.[64] Crowding results in reduced dry matter intake; poorer feed conversion efficiency; increased mounting, head resting, and aggressive behaviors; reduced play behavior; and increased physiological stress responses.[67] The shape of the pen is another consideration that should be considered in determining stocking density. As an example, cattle tend to prefer to spend their time along the perimeter of a pen rather than in the center, so crowding might be a factor even in a relatively large pen.[67,133] For dairy cattle, horses, and show animals of youth livestock projects, it is important to have the individual animal used to people working within its personal space so that there is no discomfort in being touched by a member of another species, even a stranger.

By understanding each of the various reactive distances, it is possible to minimize the stress of working with livestock. During movement, the flight distance between an individual and approacher varies. In animals that have some experience with humans, as would be the case with feedlot steers, the flight distance might be 4 or 5 feet. For range cattle it might be 100 feet or more. Social ranking also affects flight distance; higher ranking cows have a greater flight distance to humans than lower ranking ones do.[135] Using the flight distance in a controlled manner can be very efficient for handling livestock. With a slow approach, a person will notice that the animals begin to move away, keeping that specific distance as a minimum separation from the person approaching. If the person backs up a little, the movement by the animals will stop. That backward movement caused the person to step just outside the flight distance of the animals. As the person moves forward again, passing just into the flight distance, the animals will begin moving (Figure 2.11). Controlled movement continues by taking a small step directly toward the animal just into the flight distance and then stepping back again to avoid putting excessive stimulus pressure on the animals. Shifting forward and backward will result in the desired movement.

Which way the animal moves is controlled by the general position of the person relative to the animal. If a person is perpendicular to the withers and the cow is restricted by a fence on its opposite side, it could go either forward or backward when the flight distance is breached. If the handler moves into the cow's flight distance slightly behind the withers, the animal will move forward along the fence. If the person moves forward toward the head instead, the animal will move backward. Thus with a line of cattle in a raceway, moving from the front of the line toward the rear stimulates the lead cows to move forward, with the rest following. On the other hand, walking inside the flight distance from the rear toward the head of a line of cows will result in the cattle turning back toward the rear to get away. To move the line forward, it is important to

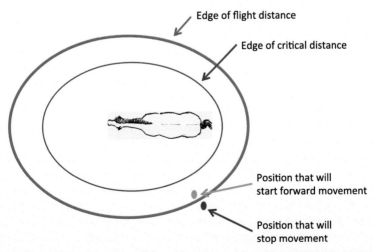

FIGURE 2.11   Flight distance is useful in moving livestock because stimuli that are applied just inside it will move the animal away. When the stimulus backs just outside the flight distance, movement by the animal will stop. © *Don Höglund (2015)*.

walk outside the flight distance when going to the front of the line so as to not disturb forward momentum. Without a fence present, the animal also has the option of angling away from the person.

The critical distance is also important to keep in mind when working cattle that are likely to attack. They will respond as described to slight stimulus pressure on the flight distance, but if the approaching human continues into the critical distance, an attack is likely. When the person remains inside the flight distance but just outside the critical distance, the animal may be upset and show obvious stress at having a person so close (as an example, a bull might paw the ground and snort), but they do not attack. Once the critical distance is breeched, however, the animal will respond aggressively. The critical distance becomes relatively insignificant for animals that have become used to being handled, and for them, the personal distance becomes important.

Goats are particularly sensitive about having individual space. Crowding will usually result in animals butting each other.[43]

Group Movement

Cattle and horse roundups provide the classic picture of moving several animals at a time. In reality, this happens in a number of other circumstances too. Forced movements, as in the roundups, and voluntary movements, as in feral horses going to a watering hole, differ in the order of individuals within the group. Cattle and sheep are very dependent on a leadership/followership relation compared to other species. The concept of leadership is distinct from dominance, with the leader of a group's movement not necessarily being the most dominant individual. Leadership is also independent from the milking order.[24]

In feral horse herds, any member of the herd can initiate movement, and the rest tend to join in because of social facilitation.[74] If a subordinate initiates this change in location, they will usually allow a higher ranking horse to pass and take the lead.[150] The stallion is typically the leader of the harem group, leading about two-thirds of the time. Stallions will also drive their harem band, usually to move away from another band or stallion, but occasionally to tighten up or direct group movements.[74,150] A lead mare takes over the remainder of the time or if the band lacks a stallion.

Cattle movement varies as to whether it is forced or voluntary. In forced movements, the lead cows tend to be low in the social order, with dominant animals in the middle of the herd.[64] In voluntary movements, the leaders are near the top in social rank, followed by middle ranking cows, with the top dominance cows near the middle and low dominance animals in the rear.[42]

When grazing on pasture, "active followership" is a strong force in how cattle move. The cow that starts movement can be any random individual. Initially, she will be followed by her close associates and close neighbors. Movements will start and stop more often if others are not following: she is adjusting her behavior to that of other herd members.[112] The animals tend to orient their bodies in the same direction as the most forward individuals, and then retain that parallel path as they move.[42] The more familiar the cattle are with each other, the stronger the active followership tendency is. Newly added cattle tend to orient randomly and away from the more established cattle.

Sheep have a very strong followership behavior and a poor leadership trait. Leaders are more apt to be older animals, more sociable ones, and of a certain breed in mixed flocks.[100] Animals that are in the rear are less tolerant of social disruptions. This seems to hold true in both voluntary and involuntary movement.[85] In feral sheep flocks, there is a fairly strict order within a line of moving sheep; however, if the leader stops, another tends to take over as the leader. Even with only two animals, when either is forced to move, the other will follow, unlike goats, which typically separate.[43] In mixed flocks, movement is more apt to be started by the goats, with the sheep following. This trait has been used by livestock truckers who bring a trained "Judas goat" that will circle through a pen of sheep and lead them directly into the truck. The goat then ducks out of the way, and when all sheep are loaded, the goat exits.

Pigs are not particularly active over long distances, with trotting being their primary gait for that travel. They tend to move as a group, but in a more random fashion than horses or cattle. In the wild, the majority of their time is spent slowly moving as they forage.[128]

LOCOMOTION

Locomotion is an important part of normal behavior for livestock. Not only is it used for finding food in pasture situations, it is associated with escape behaviors when predators are threatening. Lameness is a significant problem in dairy herds and is an underlying reason for cows being culled. Rapidity or reluctance to move can be indicators of an animal's welfare. Horses evolved to run first and then recheck. Cattle are more apt to fight a predator if a simple escape route is not obvious. Pigs have developed tonic immobility to indicate stress, especially during physical restraint. This is in addition to their increase in vocalizations.[28] Specific gaits are not particularly relevant to discussions on handling livestock; although they are of significant interest to people who ride horses. In the scenarios discussed within the later chapters on specific handling techniques, the walk is the gait of choice. In those discussions, the important consideration is the comparative speeds between the human pace than that of the animal. We also know speeds can vary depending on motivation. General descriptions indicate the human and horse walking speeds are about three to four miles per hour, and cattle walk about two to three miles per hour.

Laterality

Humans are typically right handed or left handed, and animals also show a favorite side. Exact controls of what will determine the preferred side are thought to be primarily under genetic control, with human fetuses showing a strong side bias in utero.[79] Sex, birth weight, age, and neonatal stress also seem to have an influence on laterality.

Laterality in horses is confusing, and results of different studies suggest horses have strong preferences of either left or right sides, with possible variations by breed and by sex. Foals tend to be ambidextrous, developing laterality later.[84,89] The lead favored by horses during the gallop is one measure of side preferences. As an example, when the last foot of the front pair of legs or rear pair to hit the ground during a stride is the left one, the horse is said to be in the left lead. Preferences for a lead foot used most often determine laterality. When the use of the left lead is preferred to the use of the right lead, the horse is said to have left laterality. The confusion becomes obvious in various study results. Some studies show that there is a strong preference for the right lead, being favored by up to 90% of racing Thoroughbred, Arabian, and Quarter horses.[84,152] In another study of 219 Thoroughbred and Thoroughbred cross saddle horses, 48% of the horses favored the left lead, 43% favored the right lead, and 9% were considered ambidextrous.[96] A third study showed a strong left preference by Thoroughbred and Standardbreds, but no preference by Quarter horses.[90] Sex also may play a role because male horses are significantly more left lateralized and females are more right sided,[98] which might explain the results of the race horses. Preferences for sidedness do exist but might not always be correlated between events. A strong preference for the use of a specific foot for pawing occurs in 77% of horses and for a specific foot with which to initiate walking in 67% of horses.[95] These are also used to determine laterality and can vary from each other and from lead preferences.

An interesting connection to equine laterality commonly discussed by horse enthusiasts is whether the direction of a facial hair whorl was correlated with a right or left side preference. Studies suggest that there may not be a basis for this. One study found horses with a left preference have significantly more facial whorls that were counterclockwise, and those with a right preference have clockwise whorls.[96] Another study of horses showed that a clockwise whorl was associated with horses showing a stronger left bias than do those where the hair goes in a counterclockwise direction.[127] Laterality has also been demonstrated in how a horse will use one nostril to smell the feces of a stallion. Horses under four years of age usually use their right nostril first, but they lose that preference as they get older.[89]

Laterality in cattle is generally discussed relative to preferences for sides while lying and its relationship with potential problems. Beef cows that have recently eaten and dairy cattle in freestall housing that are not lactating and are nearing calving time show a left side to the ground preference (56% of the bouts), while others do not have a side preference.[1,103,142,147] Cows do tend to alternate down sides in subsequent lying events. The tendency to lie on the right side increases with increasing age and size. An interesting correlation has been made between cows that lie more on their left side and an increased probability of a right-sided mastitis episode.[69]

Studies of laterality in sheep show that strong preferences do not exist. Neonatal lambs do not show right or left side preferences,[79] but within a few months there is a slight right bias.[145] Older ewes are also hard to evaluate. Most show less lateralization than do the older lambs, and if there is a preference, it tends to be task specific.[145] Those which have a left preference during movement tend to have a fairly strong preference for moving to the left. Those which would go to the right are less likely to choose the same direction each time.[2]

Daily Time Budget

Time budgets for livestock species vary somewhat depending on the age of the animal, availability of food and water, season of the year, and the type of housing, but generalizations can be made (Figure 2.12).[88] Complicating the comparison of time budgets between species is the variability of study methodology. Some studies report activities over a 24h period while others report data gathered only during daylight hours.

Time budgets for horses reflect that the animals spend most of the 24h period eating or standing regardless of whether they are in stalls, pastures, or range land.[7,26,32,113,149] Of the time spent standing, however, part of it is spent in dozing or in light sleep.[88] In addition, the type of housing in which weanlings are raised will affect their longer term activity behaviors.[61] When compared to weanlings raised in box stalls, those raised in small groups in paddocks will show more time moving, similar to the time budgets of a feral horse. They also show a broader range of behaviors, a stronger motivation to graze, and a stronger desire to be near other horses. Box stall-raised weanlings spend significantly more time showing aberrant behaviors such as licking the stall, kicking the stall, pawing, and bucking. These abnormal behaviors in horses can also be affected by the design of the barn and whether the horse can view other horses or the outdoors.[20]

Dairy cows will spend between 8 and 15h a day lying down.[142,144] Beef cattle show some differences when on pasture because they spend over 50% of their time during the day grazing and about three-fourths of that amount ruminating.[42,70] The time spent grazing will also tend to peak around sunrise and sunset, with rumination peaking shortly after nightfall.[42] Grazing behaviors also are associated with an increased amount of time walking, because the animal needs to travel between water and food sources and sheltered areas.

Sheep and goats on pasture share similar time budgets.[106] These animals typically spend more time walking during daylight hours and become relatively inactive at night.[105]

For pigs in pens, a large portion of the day is spent lying (about 80%) with barrows doing this slightly more often than gilts. Conversely, gilts spend about 2% more of their time standing.[40]

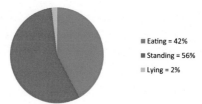

Horse 24 hour activity time budget[7, 149]

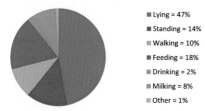

Dairy cattle 24 hour activity time budget[19, 144]

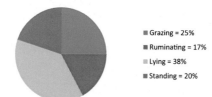

Beef cattle 24 hour activity time budget[42]

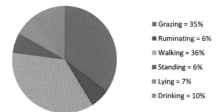

Sheep and goat 14 hour daylight activity time budget[106]

Pig 24 hour activity time budget[40]

FIGURE 2.12   The average daily time budgets for livestock species.

REFERENCES

1. Albright JL, Arave CW. *The behaviour of cattle*. New York: CAB International; 1997. pp. 306.
2. Anderson DM, Murray LW. Sheep laterality. *Laterality* March 2013;**18**(2):179–93.
3. Baldwin BA. Ability of goats and calves to distinguish between conspecific urine samples using olfaction. *Appl Anim Ethol* June 1977;**3**(2):145–50.
4. Baldwin BA, Meese GB. The ability of sheep to distinguish between conspecifics by means of olfaction. *Physiol Behav* May 1977;**18**(5):803–8.
5. Barroso FG, Alados CL, Boza J. Social hierarchy in the domestic goat: effect on food habits and production. *Appl Anim Behav Sci* August 2000;**69**(1):35–53.
6. Bartoš L, Bartošová J, Starostová L. Position of the head is not associated with changes in horse vision. *Equine Vet J* 2008;**40**(6):599–601.
7. Beaver BV. A day in the life of a horse. *Vet Med/Small Anim Clin* February 1983;**78**(2):227–8.
8. Beilharz RG, Zeeb K. Social dominance in dairy cattle. *Appl Anim Ethol* January 1982; **8**(1–2):79–97.
9. Birke L, Hockenhull J, Creighton E, Pinno L, Mee J, Mills D. Horses' responses to variation in human approach. *Appl Anim Behav Sci* October 2011;**134**(1–2):56–63.
10. Blackshaw JK, Blackshaw AW, McGlone JJ. Buller steer syndrome review. *Appl Anim Behav Sci* October 1997;**54**(2):97–108.
11. Boe KE, Faerevik G. Grouping and social preferences in calves, heifers and cows. *Appl Anim Behav Sci* February 2003;**80**(3):175–90.
12. Boissy A, LeNeindre P. Behavioral, cardiac and cortisol responses to brief peer separation and reunion in cattle. *Physiol Behav* May 1997;**61**(5):693–9.
13. Broom DM, Fraser AF. *Domestic animal behaviour and welfare*. 4th ed. Cambridge (MA): CAB International; 2007. pp. 438.
14. Camerlind I, Turner SP. The pig's nose and its role in dominance relationships and harmful behavior. *Appl Anim Behav Sci* May 2013;**145**(3–4):84–91.
15. Carroll J, Murphy CJ, Neitz M, Ver Hoeve JN, Neitz J. Photopigment basis for dichromatic color vision in the horse. *J Vis* 2001;**1**(2):80–7.
16. Christensen JW, Ladewig J, Sondergaard E, Malmkvist J. Effects of individual versus group stabling on social behavior in domestic stallions. *Appl Anim Behav Sci* January 23, 2002;**75**(3):233–48.
17. Christensen JW, Rundgren M. Predator odour *per se* does not frighten domestic horses. *Appl Anim Behav Sci* July 2008;**112**(1–2):136–45.
18. Christensen JW, Zharkikh T, Ladewig J, Yasinetskaya N. Social behavior in stallion groups (*Equus przewalskii* and *Equus caballus*) kept under natural and domestic conditions. *Appl Anim Behav Sci* February 22, 2002;**76**(1):11–20.
19. Cook NB. *Time budgets for dairy cows: how does cow comfort influence health, reproduction and productivity?* https://www.vetmed.wisc.edu/dms/fapm/publicats/proceeds/TimeBudgetsand-DairyCowsOmaha.pdf, pp. 9, downloaded March 19, 2014.
20. Cooper JJ, McDonald L, Mills DS. The effect of increasing visual horizons on stereotypic weaving: implications for the social housing of stabled horses. *Appl Anim Behav Sci* August 2000;**69**(1):67–83.
21. Crowell-Davis SL. Flehmen. *Compend Equine* 2008;**3**(2):91–4.
22. Crump D, Swigar AA, West JR, Silverstein RM, Müller-Schwarze D, Altieri R. Urine fractions that release flehmen in black-tailed deer, *Odocoileus hemionus columbianus*. *J Chem Ecol* 1984;**10**(2):203–15.

23. Des Roches ADB, Richard-Yris M-A, Henry S, Ezzaouïa M, Hausberger M. Laterality and emotions: visual laterality in the domestic horse (*Equus caballus*) differs with objects' emotional value. *Physiol Behav* June 2008;**94**(3):487–90.

24. Dickson DP, Barr GR, Wieckert DA. Social relationship of dairy cows in a feed lot. *Behaviour* 1967;**29**(2–4):195–203.

25. Dien J. Looking both ways through time: the Janus model of lateralized cognition. *Brain Cogn* August 2008;**67**(3):292–323.

26. Duncan P. Time-budgets of Camargue horses. II. Time-budgets of adult horses and weaned sub-adults. *Behaviour* 1980;**72**(1–2):26–48.

27. Dziezyc J, Taylor L, Boggess MM, Scott HM. The effect of ocular blinkers on the horses' reactions to four different visual and audible stimuli: results of a crossover trial. *Vet Ophthalmol* September 01, 2011;**14**(5):327–32.

28. Erhard HW, Mendl M, Christiansen SB. Individual differences in tonic immobility may reflect behavioural strategies. *Appl Anim Behav Sci* April 1999;**64**(1):31–46.

29. Estes RD. The role of the vomeronasal organ in mammalian reproduction. *Mammalia* 1972;**36**(3):315–41.

30. Evans KE, McGreevy PD. The distribution of ganglion cells in the equine retina and its relationship to skull morphology. *Anat Histol Embryol* April 2007;**36**(2):151–6.

31. Farmer K, Krueger K, Byrne RW. Visual laterality in the domestic horse (*Equus caballus*) interacting with humans. *Anim Cogn* March 2010;**13**(2):229–38.

32. Flannigan G, Stookey JM. Day-time time budgets of pregnant mares housed in tie stalls: a comparison of draft versus light mares. *Appl Anim Behav Sci* September 10, 2002;**78**(2–4):125–43.

33. Forrest S. Quoted from the New York Times. *Am Horse* December 2014;**17**(4):56.

34. Fraser AF. *Farm animal behaviour*. Baltimore: The Williams & Wilkins Company; 1974. pp. 196.

35. French JM, Moore GF, Perry GC, Long SE. Behavioural predictors of oestrus in domestic cattle, *Bos taurus*. *Anim Behav* December 1989;**38**(6):913–9.

36. Galina CS, Orihuela A, Rubio I. Behavioural trends affecting oestrus detection in Zebu cattle. *Anim Reprod Sci* April 1996;**42**(1–4):465–70.

37. Geisbauer G, Griebel U, Schmid A, Timney B. Brightness discrimination and neutral point testing in the horse. *Can J Zool* 2004;**82**:660–70.

38. Gieling ET, Nordquist RE, van der Staay FJ. Assessing learning and memory in pigs. *Anim Cogn* March 2011;**14**(2):151–73.

39. Gilger BC. *Equine ophthalmology*. 2nd ed. Maryland Heights (MO): Elsevier; 2010. pp. 536.

40. Gonyou HW, Chapple RP, Frank GR. Productivity, time budgets and social aspects of eating in pigs penned in groups of five or individually. *Appl Anim Behav Sci* September 1992;**34**(4):291–301.

41. Grignard L, Boissy A, Boivin X, Garel JP, LeNeindre P. The social environment influences the behavioural responses of beef cattle to handling. *Appl Anim Behav Sci* May 2000;**68**(1):1–11.

42. Hafez ESE, Bouissou MF. The behavior of cattle. In: Hafez ESE, editor. *The behaviour of domestic animals*. 3rd ed. Baltimore: The Williams and Wilkins Company; 1975. p. 203–245.

43. Hafez ESE, Cairns RB, Hulet CV, Scott JP. The behavior of sheep and goats. In: Hafez ESE, editor. *The behaviour of domestic animals*. 2nd ed. Baltimore: The Williams and Wilkins Company; 1969. p. 296–348.

44. Hafez ESE, Signoret JP. The behavior of swine. In: Hafez ESE, editor. *The behaviour of domestic animals*. 2nd ed. Baltimore: The Williams and Wilkins Company; 1969. p. 349–390.

45. Hall C. The impact of visual perception on equine learning. *Behav Processes* September 2007;**76**(1):29–33.

46. Hanggi EB. The thinking horse: cognition and perception reviewed. In: *Proceedings of the 51st Annual Convention of the American Association of Equine Practitioners, Seattle, Washington*. December 2005. p. 246–255.

47. Hanggi EB, Ingersoll JF. Stimulus discrimination by horses under scotopic conditions. *Behav Processes* September 2009;**82**(1):45–50.

48. Hanggi EB, Ingersoll JF. Lateral vision in horses: a behavioral investigation. *Behav Processes* September 2012;**91**(1):70–6.

49. Hanggi EB, Ingersoll JF, Waggoner TL. Color vision in horses (*Equus caballus*): deficiencies identified using a pseudoisochromatic plate test. *J Comp Psychol* 2007;**121**(1):65–72.

50. Hargreaves AL, Hutson GD. The effect of gentling on heart rate, flight distance and aversion of sheep to a handling procedure. *Appl Anim Behav Sci* May 1990;**26**(3):243–52.

51. Harman AM, Moore S, Hoskins R, Keller P. Horse vision and an explanation for the visual behavior originally explained by the 'ramp retina'. *Equine Vet J* September 1999;**31**(5):384–90.

52. Hart BL, Jones TOAC. Effects of castration on sexual behavior of tropical male goats. *Horm Behav* 1975;**6**:247–58.

53. Hartmann E, Keeling LJ, Rundgren M. Comparison of 3 methods for mixing unfamiliar horses (*Equus caballus*). *J Vet Behav* January–February 2011;**6**(1):39–49.

54. Heffner HE, Heffner RS. The hearing ability of horses. *Equine Pract* March 1983;**5**(3):27–32.

55. Heffner HE, Heffner RS. Sound localization in large mammals: localization of complex sounds by horses. *Behav Neurosci* June 1984;**98**(3):541–55.

56. Heffner RS, Heffner HE. Hearing in large mammals: horses (*Equus caballus*) and cattle (*Bos taurus*). *Behav Neurosci* April 1983;**97**(2):299–309.

57. Heffner RS, Heffner HE. Localization of tones by horses: use of binaural cues and the role of the superior olivary complex. *Behav Neurosci* February 1986;**100**(1):93–103.

58. Heffner RS, Heffner HE. Sound localization, use of binaural cues and the superior olivary complex in pigs. *Brain Behav Evol* April 1989;**33**(4):248–58.

59. Heffner RS, Heffner HE. Visual factors in sound localization in mammals. *J Comp Neurol* March 15, 1992;**317**(3):219–32.

60. Heffner RS, Heffner HE. Hearing in large mammals: sound-localization acuity in cattle (*Bos taurus*) and goats (*Capra hircus*). *J Comp Psychol* June 1992;**106**(2):107–13.

61. Heleski CR, Shelle AC, Nielsen BD, Zanella AJ. Influence of housing on weanling horse behavior and subsequent welfare. *Appl Anim Behav Sci* September 2002;**78**(2–4):291–302.

62. Höglund D. *Nobody's horses: the dramatic rescue of the wild herd of White Sands*. New York: Free Press; 2006. pp. 265.

63. Hothersall B, Harris P, Sörtoft L, Nicol CJ. Discrimination between conspecific odour samples in the horse (*Equus caballus*). *Appl Anim Behav Sci* August 2010;**126**(1–2):37–44.

64. Houpt KA. *Domestic animal behavior for veterinarians and animal scientists*. 3rd ed. Ames: Iowa State University Press; 1998. pp. 495.

65. Hutson GD, Ambrose TJ, Barnett JL, Tilbrook AJ. Development of a behavioural test of sensory responsiveness in the growing pig. *Appl Anim Behav Sci* February 29, 2000;**66**(3):187–202.

66. Jezierski T, Jaworski Z, Gorecka A. Effects of handling on behavior and heart rate in Konik horses: comparison of stable and forest reared youngstock. *Appl Anim Behav Sci* February 15, 1999;**62**(1):1–11.

67. Jóhannesson T, Sørensen JT. Evaluation of welfare indicators for the social environment in cattle herds. *Anim Welfare* August 2000;**9**(3):297–316.

68. Johns MA. The role of the vomeronasal system in mammalian reproductive physiology. In: Müller-Schwarze D, Silverstein RM, editors. *Chemical signals: vertebrates and aquatic invertebrates*. New York: Plenmum Press; 1980. p. 341–364.

69. Kikkers BH, Ózsvri L, van Eerdenburg FJCM, Bajcsy Á, Szenci O. The influence of laterality on mastitis incidence in dairy cattle-preliminary study. *Acta Vet Hung* June 2006;**54**(2):161–71.

70. Kilgour RJ, Uetake K, Ishiwata T, Melville GJ. The behavior of beef cattle at pasture. *Appl Anim Behav Sci* April 2012;**138**(1–2):12–7.

71. Kimura R. Mutual grooming and preferred associate relationships in a band of free-ranging horses. *Appl Anim Behav Sci* September 1998;**59**(4):265–76.

72. Koba Y, Tanida H. How do miniature pigs discriminate between people? The effect of exchanging cues between a non-handler and their familiar handler on discrimination. *Appl Anim Behav Sci* January 18, 1999;**61**(3):239–52.

73. Krueger K, Flauger B. Olfactory recognition of individual competitors by means of faeces in horses (*Equus caballus*). *Anim Cogn* March 2011;**14**(2):245–57.

74. Krueger K, Flauger B, Farmer K, Hemelrijk C. Movement initiation in groups of feral horses. *Behav Processes* March 2014;**103**:91–101.

75. Krüger K, Flauger B. Social feeding decisions in horses (*Equus caballus*). *Behav Processes* May 2008;**78**(1):76–83.

76. Ladewig J, Hart BL. Flehmen and vomeronasal organ function in male goats. *Physiol Behav* June 1980;**24**(6):1067–71.

77. Ladewig J, Price EO, Hart BL. Flehmen in male goats: role in sexual behavior. *Behav Neural Biol* November 1980;**30**(3):312–22.

78. Lagerweij E, Nelis PC, Wiegant VM, van Ree JM. The twitch in horses: a variant of acupuncture. *Science* September 14, 1984;**225**(4667):1172–4.

79. Lane A, Phillips C. A note on behavioural laterality in neonatal lambs. *Appl Anim Behav Sci* May 2004;**86**(1–2):161–7.

80. Lanier JL, Grandin T, Green RD, Avery D, McGee K. The relationship between reaction to sudden, intermittent movements and sounds and temperament. *J Anim Sci* June 2000;**78**(6):1467–74.

81. Lansade L, Pichard G, Leconte M. Sensory sensitivities: components of a horse's temperament dimension. *Appl Anim Behav Sci* December 01, 2008;**114**(3–4):534–53.

82. Larose C, Richard-Yris M-A, Hausberger M. Laterality of horses associated with emotionality in novel situations. *Laterality* 2006;**11**(4):355–67.

83. Lindberg AC. Group life. In: Keeling LJ, Gonyou HW, editors. *Social behaviour in farm animals*. New York: CABI Publishing; 2001. p. 37–58.

84. Lucidi P, Bacco G, Sticco M, Mazzoleni G, Benvenuti M, Bernabò N, et al. Assessment of motor laterality in foals and young horses (*Equus caballus*) through an analysis of *derailment* at trot. *Physiol Behav* January 17, 2013;**109**(1):8–13.

85. Lynch JJ, Hinch GN, Adams DB. *The behaviour of sheep: biological principles and implications for production*. Oxon (UK): CAB International; 1992. pp. 237.

86. Macuda T, Timney B. Luminance and chromatic discrimination in the horse (*Equus caballus*). *Behav Processes* March 01, 1999;**44**(3):301–7.

87. Macuda TJ. *Equine colour vision* [Dissertation]. 2000. http://proquest.umi.com/pqdweb?index=0&did=728862111&SrcMode=1&sid=1&Fmt=2&clientid=2945&RQT=309&VName=PQD, pp. 136, downloaded February 08, 2007.

88. McGreevy P. *Equine behavior: a guide for veterinarians and equine scientists*. New York: Saunders; 2004. pp. 369.

89. McGreevy PD, Rogers LJ. Motor and sensory laterality in thoroughbred horses. *Appl Anim Behav Sci* August 2005;**92**(4):337–52.

90. McGreevy PD, Thomson PC. Differences in motor laterality between breeds of performance horse. *Appl Anim Behav Sci* August 2006;**99**(1–2):183–90.

91. Meese GB, Conner DJ, Baldwin BA. Ability of the pig to distinguish between conspecific urine samples using olfaction. *Physiol Behav* July 1975;**15**(1):121–5.

92. Mendl M, Held S. Living in groups: an evolutionary perspective. In: Keeling LJ, Gonyou HW, editors. *Social behaviour in farm animals*. New York: CABI Publishing; 2001. p. 7–36.

93. Mogensen L, Krohn CC, Foldager J. Long-term effect of housing method during the first three months of life on human-animal relationship in female dairy cattle. *Acta Agric Scand A* August 1999;**49**(3):163–71.

94. Müller R, Schrader L. Behavioural consistency during social separation and personality in dairy cows. *Behaviour* 2005;**142**(9–10):1289–306.

95. Murphy J, Arkins S. Equine learning behavior. *Behav Processes* September 2007;**76**(1):1–13.

96. Murphy J, Arkins S. Facial hair whorls (trichoglyphs) and the incidence of motor laterality in the horse. *Behav Processes* September 2008;**79**(1):7–12.

97. Murphy J, Hall C, Arkins S. What horses and humans see: a comparative review. *Int J Zool* 2009;**2009**(Article ID 721798):1–14.

98. Murphy J, Sutherland A, Arkins S. Idiosyncratic motor laterality in the horse. *Appl Anim Behav Sci* June 2005;**91**(3–4):297–310.

99. Ollivier FJ, Samuelson DA, Brooks DE, Lewis PA, Kallberg ME, Komáromy AM. Comparative morphology of the tapetum lucidum (among selected species). *Vet Ophthalmol* January 2004;**7**(1):11–22.

100. Ortiz-Plata C, De Lucas-Tron J, Miranda-de la Lama GC. Breed identity and leadership in a mixed flock of sheep. *J Vet Behav* March–April 2012;**7**(2):94–8.

101. Patt A, Gygax L, Wechsler B, Hillmann E, Palme R, Keil NM. Behavioral and physiological reactions of goats confronted with an unfamiliar group either when alone or with two peers. *Appl Anim Behav Sci* June 2013;**146**(1–4):56–65.

102. Phillips C. *Cattle behaviour & welfare*. 2nd ed. Malden (MA): Blackwell; 2002. pp. 264.

103. Phillips CJC, Llewellyn S, Claudia A. Laterality in bovine behavior in an extensive partially suckled herd and an intensive dairy herd. *J Dairy Sci* October 2003;**86**(10):2167–3173.

104. Phillips CJC, Lomas CA. The perception of color by cattle and its influence on behavior. *J Dairy Sci* April 2001;**84**(4):807–13.

105. Piccione G, Giannetto C, Marafioti S, Casella S, Assenza A, Fazio F. Effect of different farming management on daily total locomotor activity in sheep. *J Vet Behav* July–August 2011;**6**(4):243–7.

106. Pokorná P, Hejcmanová P, Hejcman M, Pavlů V. Activity time budget patterns of sheep and goats co-grazing on semi-natural species-rich dry grassland. *Czech J Anim Sci* 2013;**58**(5):208–16.

107. Presicce GA, Brockett CC, Cheng T, Fotte RH, Rivard GF, Klemm WR. Behavioral responses of bulls kept under artificial breeding conditions to compounds presented for olfaction, taste or with topical nasal application. *Appl Anim Behav Sci* September 1993;**37**(4):273–84.

108. Price EO. *Principles & applications of domestic animal behavior*. Cambridge (MA): CAB International; 2008. pp. 332.

109. Prince JH. *Comparative anatomy of the eye*. Springfield (IL): Charles C. Thomas; 1956. pp. 418.

110. Prince JH, Diesem CD, Eglitis I, Ruskell GL. *Anatomy and histology of the eye and orbit in domestic animals*. Springfield (IL): Charles C. Thomas; 1960. pp. 307.

111. Probst JK, Hillmann E, Leiber F, Kreuzer M. Influence of gentle touching applied few weeks before slaughter on avoidance distance and slaughter stress in finishing cattle. *Appl Anim Behav Sci* February 15, 2013;**144**(1–2):14–21.

112. Ramseyer A, Thierry B, Boissy A, Dumont B. Decision-making processes in group departures of cattle. *Ethology* October 2009;**115**(10):948–57.

113. Ransom JI, Cade BS. *Quantifying equid behavior—A research ethogram for free-roaming feral horses*. U.S. Geological Survey Techniques and Methods 2-A9; 2009. pp. 23.

114. Rathore A. Order of cow entry at milking and its relationships with milk yield and consistency of the order. *Appl Anim Ethol* 1982;**8**(1–2):45–52.

115. Rehkämper G, Perrey A, Werner CW, Opfermann-Rüngeler C, Görlach A. Visual perception and stimulus orientation in cattle. *Vis Res* August 2000;**40**(18):2489–97.

116. Reinhardt V, Reinhardt A. Cohesive relationships in a cattle herd (*Bos indicus*). *Behaviour* 1981;**77**(3):121–50.

117. Rekwot PI, Ogwu D, Oyedipe EO, Sekoni VO. The role of pheromones and biostimulation in animal reproduction. *Anim Reprod Sci* March 30, 2001;**65**(3–4):157–70.

118. Riol JA, Sanchez JM, Eguren VG, Gaudioso VR. Colour perception in fighting cattle. *Appl Anim Behav Sci* June 1989;**23**(3):199–206.

119. Robins A, Phillips C. Lateralised visual processing in domestic cattle herds responding to novel and familiar stimuli. *Laterality* 2010;**15**(5):514–34.

120. Roth LSV, Balkenius A, Kelber A. Colour perception in a dichromat. *J Exp Biol* August 15, 2007;**210**(16):2795–800.

121. Roth LSV, Balkenius A, Kelber A. The absolute threshold of colour vision in the horse. *PLoS One* November 2008;**3**(11):e3711.

122. Rushen J, Taylor AA, de Passillé AM. Domestic animals' fear of humans and its effect on their welfare. *Appl Anim Behav Sci* December 1999;**65**(3):285–303.

123. Salazar I, Quinteiro PS, Cifuentes JM. The soft-tissue components of the vomeronasal organ in pigs, cows and horses. *Anat Hist Embryol* September 1997;**26**(3):179–86.

124. Sankey C, Henry S, Clouare C, Richard-Yris M-A, Hausberger M. Asymmetry of behavioral responses to a human approach in young naïve vs. trained horses. *Physiol Behav* September 01, 2011;**104**(3):464–8.

125. Saslow CA. Understanding the perceptual world of horses. *Appl Anim Behav Sci* September 10, 2002;**78**(2–4):209–24.

126. Sato S. Leadership during actual grazing in a small herd of cattle. *Appl Anim Ethol* January 1982;**8**(1–2):53–65.

127. Savin H, Randle H. The relationship between facial whorl characteristics and laterality exhibited in horses. *J Vet Behav* September/October 2011;**5**:295–6.

128. Signoret JKP, Baldwin BA, Fraser D, Hafez ESE. The behavior of swine. In: Hafez ESE, editor. *The behaviour of domestic animals*. 3rd ed. Baltimore: The Williams and Wilkins Company; 1975. p. 295–329.

129. Sigurjónsdóttir H, van Dierendonck MC, Snorrason S, Thórhallsdóttir AG. Social relationships in a group of horses without a mature stallion. *Behaviour* 2003;**140**(6):783–804.

130. Søndergaard E, Halekoh U. Young horses' reactions to humans in relation to handling and social environment. *Appl Anim Behav Sci* December 22, 2003;**84**(4):265–80.

131. Stahlbaum CC, Houpt KA. The role of the flehmen response in the behavioral repertoire of the stallion. *Physiol Behav* June 1989;**45**(6):1207–14.

132. Stricklin WR, Gonyou HW. Dominance and eating behavior of beef cattle fed from a single stall. *Appl Anim Ethol* April 1981;**7**(2):135–40.

133. Stricklin WR, Graves HB, Wilson LL. Some theoretical and observed relationships of fixed and portable spacing behavior of animals. *Appl Anim Ethol* July 1979;**5**(3):201–14.

134. Stricklin WR, Graves HB, Wilson LL, Singh RK. Social organization among young beef cattle in confinement. *Appl Anim Ethol* July 1980;**6**(3):211–9.

135. Syme GJ, Syme LA. *Social structure in farm animals*. New York: Elsevier; 1979. pp. 200.

136. Takeda K-i, Sato S, Sugawara K. The number of farm mates influences social and maintenance behaviours of Japanese Black cows in a communal pasture. *Appl Anim Behav Sci* April 03, 2000;**67**(3):181–92.

137. Takeda K-i, Sato S, Sugawara K. Familiarity and group size affect emotional stress in Japanese Black heifers. *Appl Anim Behav Sci* June 03, 2003;**82**(1):1–11.

138. Takeda K-i, Sato S, Sugawara K. Affiliative group size influences grazing and recumbency behaviors in heifers immediately after beginning grazing and in high-density grazing situations. *Grassl Sci* December 2008;**54**(4):197–202.

139. Tallet C, Brilloüet A, Meunier-Salaün M-C, Paulmier V, Guérin C, Prunier A. Effects of neonatal castration on social behavior, human-animal relationship and feeding activity in finishing pigs reared in a conventional or an enriched housing. *Appl Anim Behav Sci* May 2013;**145**(3–4):70–83.

140. Timney B, Keil K. Visual acuity in the horse. *Vis Res* December 1992;**32**(12):2289–93.

141. Timney B, Macuda T. Vision and hearing in horses. *J Am Vet Med Assoc* May 15, 2001; **218**(10):1567–74.

142. Tucker CB, Cox NR, Weary DM, Špinka M. Laterality of lying behavior in dairy cattle. *Appl Anim Behav Sci* September 2009;**120**(3–4):125–31.

143. Turner SP, Nath M, Horgan GW, Edwards SA. Measuring chronic social tension in groups of growing pigs using inter-individual distances. *Appl Anim Behav Sci* June 2013;**146**(1–4):26–36.

144. Uzal S, Ugurlu N. The dairy cattle behaviors and time budget and barn area usage in freestall housing. *J Anim Vet Adv* 2010;**9**(2):248–54.

145. Versace E, Morgante M, Pulina G, Vallortigara G. Behavioural lateralization in sheep (*Ovis aries*). *Behav Brain Res* November 22, 2007;**184**(1):72–80.

146. von Borstel UU, Duncan IJH, Shoveller AK, Merkies K, Keeling LJ, Millman ST. Impact of riding in a coercively obtained Rollkur posture on welfare and fear of performance horses. *Appl Anim Behav Sci* January 31, 2009;**116**(2–4):228–36.

147. Wagnon KA, Rollins WC. Bovine laterality. *J Anim Sci* 1972;**35**(2):486–8.

148. Waiblinger S, Menke C. Influence of herd size on human-cow relationships. *Anthrozoös* 1999;**12**(4):240–7.

149. Waring GH. *Horse behavior*. 2nd ed. Norwich (NY): Noyes Publications; 2003. pp. 442.

150. Waring GH, Wierzbowski S, Hafez ESE. The behavior of horses. In: Hafez ESE, editor. *The behaviour of domestic animals*. 3rd ed. Baltimore: The Williams and Wilkins Company; 1975. p. 330–369.

151. Weeks JW, Crowell-Davis SL, Caudle AB, Heusner GL. Aggression and social spacing in light horse (*Equus caballus*) mares and foals. *Appl Anim Behav Sci* July 2000;**68**(4):319–37.

152. Williams DE, Norris BJ. Laterality in stride pattern preferences in racehorses. *Anim Behav* October 2007;**74**(4):941–50.

153. Wilson WJ, Mills PC, Dzulkarnain AA. Use of BAER to identify loss of auditory function in older horses. *Aust Vet J* March 2011;**89**(3):73–6.

154. Wolski TR, Houpt KA, Aronson R. The role of the senses in mare-foal recognition. *Appl Anim Ethol* April 1980;**6**(2):121–38.

155. Yajuvendra S, Lathwal SS, Rajput N, Raja TV, Gupta AK, Mohanty TK, et al. Effective and accurate discrimination of individual dairy cattle through acoustic sensing. *Appl Anim Behav Sci* June 2013;**146**(1–4):11–8.

156. Yeon SC. Acoustic communication in the domestic horse (*Equus caballus*). *J Vet Behav* May/June 2012;**7**(3):179–85.

157. Zanella AJ, Brunner P, Unshelm J, Mendl MT, Broom DM. The relationship between housing and social rank on cortisol, β-endorphin and dynorphin (1-13) secretion in sows. *Appl Anim Behav Sci* August 1998;**59**(1–3):1–10.

Chapter 3

Genetics, Environments, and Hormones

… the genetic lottery may determine the cards in your deck, but experience deals the hand you can play.

Thomas Lewis, *A General Theory of Love*

The wild horses that roamed the White Sands Missile Range had been free-living for several generations, interbreeding occasionally with local mares and perfecting the instincts they needed for survival. The severe drought of 1994 changed that. Removing the horses from White Sands was not going to be an easy task, but it was one contracted to Don Höglund to accomplish. Don easily spotted the poor condition of many of the surviving horses, but he also noticed something else. The original horses brought to North America by the Spanish were small horses with the "tiger striping" on the legs, as typical of the original Spanish Barb lines. The horses he was seeing on the hilly range land were predominantly huge and had feathering on the fetlocks, typical of draft horse ancestry. It became obvious that the longer legs had been genetically favored because they were needed to travel great distances in search of the scant grass in this part of the West. Whoever said it first got it right: the only real control any horse has is in the way he reacts to his environment.

Nature versus nurture—the age-old argument of what is most important in the development of an individual is no longer defined as a simple answer one way or the other. Researchers have spent numerous hours trying to learn what role our genes play in human personality development and what the significance of environmental influence is. In humans, identical twins that were separated at birth were favorite subjects to try to tease out answers, only to ultimately show how complicated the entire issue is.[26] In animals, personality profiles attempt to define patterns of behavior in similar environments. The tightness of the genetic pools can be more controlled and selective breeding has been used for many years to improve characteristics like milk production, fleece quality, and athletic performance. However, heritability of any single trait is rarely 100% guaranteed. Even clones are not identical to each other, as seen in four genetically identical mule clones with very different personalities. As we learn more via genetic profiling, we are beginning to understand

Efficient Livestock Handling. http://dx.doi.org/10.1016/B978-0-12-418670-5.00003-2

the interrelation of various traits based on their position on the genes and how proximity can affect other traits. As an example, the trait for the fine wool of the Merino sheep is located very close to the trait for excessive skin folds, making it hard to breed out wrinkles without adversely affecting wool quality. Modern science is finding newer ways to look at the genome and how it influences personality and production. The full implication of these findings for each species is not yet known.

GENES AND INDIVIDUALS

Ever since Gregor Mendel studied the laws of genetics in the mid-1800s, animal breeding has taken advantage of the concepts of selection for desirable characteristics to improve the next generations. Over time, dairy cows produced more milk and beef cattle produced more meat, first with lots of marbling and then with less. Horses became more specialized, sheep produced thick wool coats, and some goats became dairy animals. The nagging problems are that the outcomes of any one breeding are not always what is expected, and the amount of progressive changes seem limited.

Heritability is given a number ranging from 0.0, meaning there is no genetic contribution to the trait, up to 1.0, where inheritance is the only reason for the trait. (A percent figure is sometimes used instead: 0–100%.) As an example, a heritability score of 0.29 for racing Standardbreds[53] means that about 29% of the speed may be attributable in some way to the genetics of the individual. It does not mean that 29% of a racing Standardbred's speed is due to his genes and 71% due to the environment.[45] In racing Thoroughbreds, the heritability of speed is even worse. The value attributed to inheritance is less than 0.20, with the number being somewhat higher for short races and lower as the distance increases.[45] Researchers have also looked at traits that could influence the interaction of animals with humans. One example is that gilts have a moderate heritably of 0.376 for a fear of humans.[21]

Genetically, an animal gets 50% of its genes from each parent, so the physical and behavioral traits should be a half and half blend of both. But it is not that simple. In a study of dog behavior, pictures of the first-generation offspring of a Cocker Spaniel and Basenji show puppies that are similar to each other: they look like one would expect for such a cross. If the first-generation Cocker×Basenji puppies are then crossed to other first-generation Cocker×Basenji crosses, the second generation is extremely variable in appearance, ranging from very Cocker-like to very Basenji-like. The researchers reported the same extreme variations occurred in personality types of the second generation.[51] Some genetic traits are dominant and, if inherited from one parent, will be obvious in the offspring.

The heritability of different traits is variable by trait, breed, and species. A study comparing German Angus and Simmental cattle can serve as an example of this.[12] While vocalization during restraint is similar between the two breeds,

Simmental are significantly more likely to urinate or defecate during restraint. They also tend to run rather than walk when being worked. As a result, the heritability factor for temperament scores is 0.0–0.59 for Simmental, compared to 0.0–0.61 for German Angus. While this might not seem significant, when factored with weigh gain, it becomes important. Brahman cattle are more excitable than English breeds, but if repeatedly prodded with a hot shot, Brahman are likely to lie down and become immobile. This behavior is rare in English breed cattle.[17] In horse racing, Thoroughbreds and Standardbreds excel, while Arabians stand out in endurance challenges. With pigs, there are breed differences in the speed of movement, and certain hybrid lines of high producing sows are difficult to drive.[17] The ryanodine receptor 1 (skeletal) or RYR1 gene in swine is associated with increased blood lactate and creatine phosphokinase (CPK) values, indicative of an increased stress response, compared to pigs that do not carry the gene.[18] Sheep behaviors can vary by genotypes too.[13,43]

There are negatives as well as positives associated with artificial selection for specific traits. By placing emphasis on certain traits, it means that others become less important. As a result, those ignored traits can be lost or modified in undesirable ways. They may also be desired in one sex but considered negative in the other. Selection for fine wool in Merino sheep, as an example, has resulted in diminished maternal behavior in the ewes.[35] In the end, there are few physical or behavior traits that are 100% heritable.

Temperament

Temperament and personality are difficult topics to discuss because there are no universal definitions for the terms. This is made even more complicated because there is now evidence that personalities mature as the result of interactions with the environment.[34] To describe a personality, humans tend to use broad characteristics in an attempt to explain what the animal may be feeling and thinking.[15,46] Temperament also is a human construct that is expressed as the foundation of personality,[34] and animal researchers agree even less about how to define temperament.[5,14] While it is more accurate to discuss patterns of behavior instead of "personality" or "temperament," the literature commonly uses these terms, and it is the literature that is the basis for the following discussion.

Temperament is generally described as a trait that plays a big role in how an individual animal interprets and reacts to its environment, external stimuli, and handling. Heritability of temperaments has been shown to some extent in livestock species, but there is genetic variability.[12] This might partially be connected to the relationship of temperaments with external appearances. For example, the higher on the head facial hair whorls were located in beef cattle, the greater was the tendency for restlessness in squeeze chutes. However, the consistency of this finding and the importance of it in selective breeding is yet to be understood.[40] Reactivity to sensory inputs is another factor that influences temperament scoring, with or without obvious genetic influences. In cattle, excitable temperaments are associated with poorer daily weight gain, reduced pregnancy rates, tougher meat, and a higher incidence of dark cutters compared to calm temperaments.[12,29,57] Holsteins have been shown to be more sound and touch sensitive than are beef cattle, indicating a genetic connection, and motion-sensitive cattle do poorer on temperament testing.[27] Cattle that are more easily agitated when handled are also more likely to startle to sudden sounds or movements.[27] Mismothering by ewes is associated with poor temperaments.[29]

Genomic studies of temperament have increased over time and are now providing a better understanding of what genes carry certain traits.[19] Until a time comes where new knowledge can be used in highly specialized breeding programs, it remains important to understand the relationship of temperaments and production.

Some validity is shown in various ways "personality" is interpreted, such as in the identification of "flighty."[15,36] Fortunately, that happens to be a characteristic of concern in livestock species. It is also the trait that was particularly important for the survival of individuals and species. Flightiness as an individual's trait has proven to be consistent over time.[56] Most animals are more reactive to certain types of stimuli, such as a horse to fluttering plastic caught on a fence or the odor of a rotting carcass. Reactivity to a certain type of sensory stimulus is consistent over time and does not predict reactivity to other types of sensory stimuli.[28] Reactivity in horses to novel objects is related to nervousness when being ridden, and half siblings tend to behave the same way.[30,58] A recent study suggests that part of the reactivity might actually be related to vision, at least in horses. Breed differences were found in the approximately one-third of adult horses that were either near- or farsighted.[1] Warmbloods and Shires tended toward being farsighted and Thoroughbred crosses were more likely to be nearsighted.

A nasty disposition in certain lines of racehorses is an example of the genetic relationship to personality. The behavior is tolerated because of the ability of the animals to excel in racing. They are just extremely difficult to handle, and the handling used may also accentuate the nasty disposition. Add to this mix of variability the influence of the dam on the offspring in the postnatal period—they learn behavior from her. It is understandable that physical as well as behavioral features are not always highly influenced by genetics.

An interesting discussion goes on in several species about the relationship between an individual's hair coat color and its personality. This connection goes back to embryology, because a part of the early embryo that gives rise to the nervous system also gives rise to the skin. Calico cats are often considered to be aggressive, and gray horses are docile. Thinking about survival of light colored horses in the wild should bring thought to such generalizations. A predator should be able to see a light colored horse easier, so being white or gray should be less desirable and survivors should be more flighty. People tend to remember the animals that fit the stereotype and forget those that do not. While generalizations are probably unfounded, occasional studies do show significance. One such study has shown that fear-like reactions in Icelandic horses are significantly greater in those carrying the Silver mutation than in horses of other colors. A follow-up to that study has shown that sires with this trait passed the reactivity on to their progeny regardless of the foal's color.

Role of Culling

The culling of individuals from a herd is usually done based on the lack of performance or on undesirable physical traits, such as lameness, small stature, poor conformation, or malformations. Temperament should also be considered. The trait of becoming easily and excessively agitated is persistent over time. Aggression, such as that seen with the routine mixing of pigs, poses the potential for serious problems. These animals are dangerous to handlers and other animals, as well as being highly distressed, so culling them from a herd and breeding program becomes a safety factor.[16,44,54] Culling for temperament and selections against bad traits is particularly important when considering which males to use in a breeding program because one individual will pass genes on to many more offspring than will the female of the species.

ROLE OF THE ENVIRONMENT

In the nature versus nurture discussions, it is hard to tease out the significance of the environment in how an animal responds to various stimuli. The breadth of this topic ranges from the interactions with the dam to a newer discipline called epigenetics. Domestication has changed livestock species to meet environmental needs and human desires. In several horse breeds, it is desirable for foals to be born as close to January first as possible for the competitive advantage of a larger size. Since natural ovarian cycles begin in April or May, mares are often kept under lights to encourage earlier ovarian cycling. That too tends to influence genetic selection for mares that cycle earlier.

All of the species under discussion are typically kept in groups that are significantly larger than would occur in the wild. Other environmental factors impact welfare, physiology, and social structures too. To minimize aggression to humans and within the herds and flocks, males are typically castrated. As a result, a whole new dimension of how castrated animals fit into a group has been introduced. Assisted births associated with dystocias in cattle and sheep are related to high levels of physical damage and a reduction in the calf or lamb's ability to thrive.[8,32] Heat or cold reduce estrus cycle length and the intensity of estrous behavior, as well as the duration of eating behavior and general activity.[41] A significantly greater amount of bruising of meat is correlated with the use of electric prods or plastic pipe to move cattle and with cattle that have gone through an auction barn rather than being shipped directly from their farms of origin.[25] Pigs that are negatively handled on their way to and at slaughter facilities will have higher cortisol, lactate, and CPK levels, and meat quality is poorer when compared to that from pigs that receive minimal handling.[3,6]

The environment can positively impact welfare of livestock too. Stabled horses that have some access to herdmates, multiple feedings a day, and sensory stimulation show fewer aberrant behavior. They are also more alert and will sleep more in lateral recumbency than those that are stabled without enrichments.[55] From a production standpoint, it is important to note that enriched environments do not negatively impact meat quality.[24] Enrichment opportunities are numerous. Horses might get large balls to play with, or they might get to spend part of their day in a pasture or paddock with other horses instead of remaining stalled. Cattle are naturally curious so occasionally adding and removing things encourages exploration. Pigs need objects to root.

Socialization

Two types of social interactions are important for all young animals. The first is *imprinting*, a sensitive time period when individuals learn who their mother is and what their own species is. The second is *socialization*. That involves an individual learning to accept close proximity with members of another species. In all species, there is a sensitive period in which these processes must occur. Exact time periods have only been worked out for dogs (3–12 weeks of age)[51]; however, in species that are born with their eyes and ears open and that are capable to running with their dam soon after birth, the socialization period is much earlier. For foals and calves, imprinting occurs within the first few days after birth, and socialization probably happens within the first week or two. Lambs, kids, and piglets are not quite as developed at birth so their imprinting and socialization periods may start a few days to a week or so later.[49]

The proper introduction of young livestock to humans, especially during the socialization period, can have significant benefits in the long run. Regular, gentle handling by humans results in lowered stress responses, including lowered heart rates and cortisol levels.[49] Heifers will continue eating when humans are present and show improved behaviors during the first milking. Sheep are easier to handle and dairy goats have less reduced residual milk.[49]

There has been a great deal of discussion about "imprinting" within the popular equine literature, but the term is typically used incorrectly. Imprinting is an introduction to the foal's own species. Socialization is its introduction to people, dogs, cattle, and other species. In the popular literature, "imprinting" actually describes habituation of a foal toward a variety of potentially fear-inducing stimuli, such as clipper noise and foot handling.

Stress

Stressors can be significant factors that livestock encounter during handling, particularly for those animals that were poorly socialized to humans. While a concise definition of stress is elusive, it may be helpful to think of *stressors* as anything that disrupts natural physiological balance and a *stress response* as the body's adaptations to restore homeostatic balance.[42,50] When stressors are great enough or occur for a long enough period, an animal may lose its ability to adapt, and the resulting state is called *distress*. Early work by Hans Seyle helped demonstrate the existence of biological stress responses by showing that a wide variety of noxious stimuli caused a very consistent set of pathologic changes in laboratory rats. Research has confirmed similar findings in other animals too.

From an animal's perspective, a human has a propensity for quick, unpredictable movements, loud noise, and rough handling—common causes of chronic stress responses. Previous negative experiences with people will make animals more standoffish, and the mere presence of a person becomes stressful.

In the last few years, undefined phrases such as "low-stress" and "cow-comfort" have become popular in the livestock husbandry. That said, it is important to not forget that positive contributions to the perception of humane production demand precise and unambiguous language. People evaluating low-stress handling techniques have raised the questions of what exactly stress is, and how do we determine if it is "low" or "high"? If you ask 12 people to define "stress" you would likely get 12 different answers. This creates an interesting challenge when attempting to determine the level of animal stress on a particular farm and whether the stress level is "low" or "high." If we struggle to define stress, how can we measure it?

Stressors can be described by their characteristics, such as duration, frequency, intensity, predictability, and the ability to be controlled. It is important to note that while stressors can be physical things, such as heat, cold, and starvation, psychological factors can also trigger the stress response in an animal in the absence of any physical threat. Stress responses evolved as an adaptive survival mechanism for animals, but they come with a "biological cost" to the animal. Stress does not make an animal sick, but it weakens the immune system's ability to respond.

One of the best measures of stress in livestock is how long it takes the animal to approach a human. Blood indicators provide other measures of stress and distress. As an example, when pigs are handled by humans, cortisol, lactate, and CPK are significantly higher than baseline levels.[18,33] These measures are not without issues. For example, not all stressors cause changes in cortisol values. Obtaining a blood sample in itself can be stressful, especially in wildlife or zoo animals, and not all individuals respond in the same way to a potential stressor.

In dairies, the stock person can be a significant cause of stress either by lower levels of technical competence, abrupt actions, loud voices, or changes in personnel.[49] In these stressful situations, milk yield is often negatively affected. Cows take longer to enter the milking parlor, and they defecate more often in the parlor. There is also a higher incidence of lameness.[49] For sheep, chronic stress can result from rough handling, sheepdogs, novel environments, social mixing, and lameness. These can directly affect reproductive function, impair body and wool growth, reduce meat quality and immune function, and alter activity and feeding behaviors.[9,37] In swine, the farrowing rate and total number of piglets born is reduced with rough handling, in addition to growth rate and age of first estrus.[20,22,49] In contrast to these negatives, positive changes in handling techniques have been shown to improve productivity.[23]

Epigenetics

Epigenetics is a relative new field of study that looks at how the environment affects gene expression. We do not understand the full implications of this field on domestic animals, but it is reasonable to think that there will be some overlap between what is learned in other species and what happens in livestock.

We now can also appreciate that stress not only affects the individual undergoing the stress, but also it impacts fetuses of pregnant animals. In normal individuals, the hypothalamus area of the brain responds to increased stress by releasing the *corticotropin releasing hormone* (CRH) (Figure 3.1). This is carried in the blood stream to the pituitary gland at the base of the brain and causes the pituitary to release the *adrenocorticotropic hormone* (ACTH). The

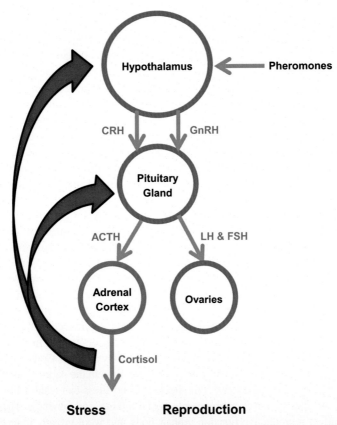

FIGURE 3.1 The hypothalamic-pituitary-adrenal (HPA) axis on the left side is important in stress, and the interactions of pheromones on the ovaries (right side) share similar parts of the brain and pituitary gland. CRH, corticotropin releasing hormone; ACTH, adrenocorticotropic hormone; GnRH, gonadotropin releasing hormone; LH, luteinizing hormone; FSH, follicle stimulating hormone; red arrows show the feedback loops.

ACTH is then carried via the blood to the adrenal glands in the abdomen. These are the glands that release *cortisol*, the stress hormone. Cortisol is the hormone commonly measured to determine if the animal is stressed. While cortisol is not always the perfect measure of stress, when present in higher than normal levels, there is a strong indication of chronic stress present. Increased levels of cortisol trigger a feedback loop to both the hypothalamus and pituitary gland that causes them to lower their hormone production.

If the animal is pregnant, the fetus she carries is also subjected to the hormone fluctuations. As a result, the fetal adrenal glands will be affected, making them more susceptible to stress activation. These offspring are predisposed to being excessively reactive, being hyperactive, and having attention disorders. Because the females of most of the livestock species are almost continually pregnant, the implications are significant relative to how well their offspring react to handling. If the young are more sensitive to stressors, they are more likely to be stressed during pregnancy and continue the problem on to another generation.

In addition to fetuses being affected by cortisol levels in the dam, there is evidence in rodents that a very stressful experience to a female before she is pregnant can have negative effects on her offspring when she does become pregnant and can even affect second pregnancy offspring.[10,31,52] We do not yet know if this is true in other species, but the potential should be of concern.

Males are not immune from these findings either. If they are a fetus at the time of the stress, they are subject to the same problems of reactivity, hyper behavior, and attention disorders. It is now known that their sperm will be changed in a way that will affect the brain of their offspring to cause them to also have abnormal stress responses.[11,48]

INTERACTION OF GENETICS AND THE ENVIRONMENT

Ease of handling results from a combination of genetics of things, such as temperament and of the environment, including early learning. Pigs serve as a good example of this complicated interaction. The genetic control of aggression in pigs is well studied, but it has been shown that those born into larger litters also become more aggressive.[7] Different breeds of pig also respond to the environment differently. Chinese Meishan pigs are known to have a hyperactive hypothalamic-pituitary-adrenal (HPA) axis with relatively high levels of circulating cortisol, but they also have a low reactivity to stressors compared to European breeds.[38]

The selection of the appropriate genotypes for a particular situation facilitates the ease of handling, lower stress, and the reduction of aggression. In some cases, a parent can produce offspring that are very protective if kept under range conditions but are easy to handle in confinement. At the opposite extreme are animals like Romanov ewes, which if kept in intensive systems show good maternal skills but are easily disturbed in the presence of humans in environments where humans are infrequently present.[29]

ENDOCRINE INFLUENCES

Reproduction in livestock is critically tied to the role of hormone physiology. This involves a complex interaction of inputs and interrelationships that are not particularly relevant to handling. There is one aspect that does tie in to previous discussions of pheromones in Chapter 2. These special chemicals communicate with one section of the brain called the hypothalamus. It is in the hypothalamus that the nervous system interacts with a variety of inputs, such as daylight photoperiods, stress, smells, presence of a male, and pheromones and with feedback loops from the pituitary and adrenal glands. In females, the hypothalamus releases a hormone (gonadotropin releasing hormone or GnRH) that goes to the pituitary gland; the pituitary in turn releases the luteinizing hormone (LH) and the follicle stimulating hormone (FSH) that work on the ovary for successful reproduction (Figure 3.1). The priming function of pheromones and other inputs to the hypothalamus are known to hasten sexual maturity, induce ovulation, and reduce postpartum anestrus in cattle, sheep, goats, and pigs.[2,47] There are also differences in other parts of the brain of females that are under the influence of estrogen. Examples include olfactory and visual processing regions, which are markedly more sensitive to processing sensory cues given by males.[39]

REFERENCES

1. Bracun A, Ellis AD, Hall C. A retinoscopic survey of 333 horses and ponies in the UK. *Vet Ophthalmol* 2014;**17**(S1):90–6.
2. Brooks PH, Cole DJA. The effect of the presence of a boar on the attainment of puberty in gilts. *J Reprod Fertil* December 1970;**23**(3):435–40.
3. Brown SN, Warriss PD, Nute GR, Edwards JE, Knowles TG. Meat quality in pigs subjected to minimal preslaughter stress. *Meat Sci* July 1998;**49**(3):257–65.
5. Buss AH. *Personality: temperament, social behavior, and the self.* Needham Heights (MA): Allyn & Bacon; May 1995. p. 420.
6. Channon HA, Payne AM, Warner RD. Halothane genotype, pre-slaughter handling and stunning method all influence pork quality. *Meat Sci* November 2000;**56**(3):291–9.
7. D'Eath RB, Lawrence AB. Early life predictors of the development of aggressive behavior in the domestic pig. *Anim Behav* March 2004;**67**(3):501–9.
8. Dwyer CM. Birth difficulty effects on mother-offspring behavior and offspring development. In: *4th Boehringer Ingelheim expert forum on farm animal well-being.* May 27, 2011. p. 41–44.
9. Dwyer CM, Bornett HLI. Chronic stress in sheep: assessment tools and their use in different management conditions. *Anim Welfare* August 2004;**13**(3):293–304.
10. Franklin TB, Mansuy IM. Epigenetic inheritance in mammals: evidence for the impact of adverse environmental effects. *Neurobiol Dis* July 2010;**39**(1):61–5.
11. Franklin TB, Russig H, Weiss IC, Gräff J, Linder N, Michalon A, et al. Epigenetic transmission of the impact of early stress across generations. *Biol Psychiatry* September 01, 2010;**68**(5):408–15.
12. Gauly M, Mathiak H, Hoffmann K, Kraus M, Erhardt G. Estimating genetic variability in temperamental traits in German Angus and Simmental cattle. *Appl Anim Behav Sci* October 10, 2001;**74**(2):109–19.

13. Goddard PJ, Fawcett AR, Macdonald AJ, Reid HW. The behavioural, physiological and immunological responses of lambs from two rearing systems and two genotypes to exposure to humans. *Appl Anim Behav Sci* March 2000;**66**(4):305–21.

14. Goldsmith HH, Buss AH, Plomin R, Rothbart MK, Thomas A, Chess S, et al. Roundtable: what is temperament? Four approaches. *Child Dev* April 1987;**58**(2):505–29.

15. Gosling SD. From mice to men: what we can learn about personality from animal research. *Psychol Bull* 2001;**126**(1):45–86.

16. Grandin T. Behavioral agitation during handling of cattle is persistent over time. *Appl Anim Behav Sci* March 1993;**36**(1):1–9.

17. Grandin T. *Behavioral principles of livestock handling*. http://www.grandin.com/references/new.corral.html, downloaded 6/21/2013.

18. Guàrdia MD, Estany J, Álvarez-Rodríquez J, Manteca X, Tor M, Oliver MA, et al. A field assessment of the effect of pre-slaughter conditions and genetic-stress susceptibility on blood welfare indicators in pigs. *Anim Welfare* November 2012;**21**(4):517–26.

19. Gutirrez-Gil B, Ball N, Burton D, Haskell M, Williams JL, Wiener P. Identification of quantitative trait loci affecting cattle temperament. *J Hered* May 21, 2008;**99**(6):629–38.

20. Hemsworth PH, Barnett JL, Coleman GJ, Hansen C. A study of the relationships between the attitudinal and behavioural profiles of stockpersons and the level of fear of humans and reproductive performance of commercial pigs. *Appl Anim Behav Sci* July 1989;**23**(4):301–14.

21. Hemsworth PH, Barnett JL, Treacy D, Madgwick P. The heritability of the trait fear of humans and the association between this trait and subsequent reproductive performance of gilts. *Appl Anim Behav Sci* January 1990;**25**(1–2):85–95.

22. Hemsworth PH, Brand A, Willems P. The behavioural response of sows to the presence of human beings and its relation to productivity. *Livest Prod Sci* February 1981;**8**(1):67–74.

23. Hemsworth PH, Coleman GJ, Barnett JL. Improving the attitude and behavior of stockpersons towards pigs and the consequences on the behaviour and reproductive performance of commercial pigs. *Appl Anim Behav Sci* March 1994;**39**(3–4):349–62.

24. Hill JD, McGlone JJ, Fullwood SD, Miller MF. Environmental enrichment influences on pig behavior, performance and meat quality. *Appl Anim Behav Sci* April 1998;**57**(1):51–68.

25. Jarvis AM, Selkirk L, Cockram MS. The influence of source, sex class and pre-slaughter handling on the bruising of cattle at two slaughterhouses. *Livest Prod Sci* September 1995;**43**(3):215–24.

26. Krueger RF, South S, Johnson W, Iacono W. The heritability of personality is not always 50%: gene-environment interactions and correlations between personality and parenting. *J Pers* December 2008;**76**(6):1485–521.

27. Lanier JL, Grandin T, Green RD, Avery D, McGee K. The relationship between reaction to sudden, intermittent movements and sounds and temperament. *J Anim Sci* June 2000;**78**(6):1467–74.

28. Lansade L, Pichard G, Leconte M. Sensory sensitivities: components of a horse's temperament dimension. *Appl Anim Behav Sci* December 01, 2008;**114**(3–4):534–53.

29. Le Neindre P, Boivin X, Boissy A. Handling of extensively kept animals. *Appl Anim Behav Sci* August 01, 1996;**49**(1):73–81.

30. Le Scolan N, Hausberger M, Wolff A. Stability over situations in temperamental traits of horses as revealed by experimental and scoring approaches. *Behav Processes* December 1997;**41**(3):257–66.

31. Leshem M, Schulkin J. Transgenerational effects of infantile adversity and enrichment in male and female rats. *Dev Psychobiol* March 2011;**54**(2):169–86.

32. Leslie KE, Murray C. Impact of dystocia on newborn calf vigor. In: *4th Boehringer Ingelheim expert forum on farm animal well-being*. May 27, 2011. p. 33–39.

33. Marhcant-Forde JN, Matthews DL, Poletto R, Mccain RR, Mann DD, Degraw RT, et al. Plasma cortisol and noradrenalin concentrations in pigs: automated sampling of freely moving pigs housed in the PigTurn® versus manually sampled and restrained pigs. *Anim Welfare* May 2012;**21**(2):197–205.

34. McCrae RR, Costa Jr PT, Ostendorf F, Angleitner A, Hřebíčková M, Avia MD, et al. Nature over nurture: temperament, personality, and life span development. *J Pers Soc Psychol* January 01, 2000;**78**(1):173–86.

35. McPherson FJ, Chenoweth PJ. Mammalian sexual dimorphism. *Anim Reprod Sci* April 2012;**131**(3–4):109–22.

36. Mills DS. Personality and individual differences in the horse, their significance, use and measurement. *Equine Vet J* November 1998;**30**(S27):10–3.

37. Miranda-de la Lama GC, Villarroel M, María GA. Behavioural and physiological profiles following exposure to novel environment and social mixing in lambs. *Small Rumin Res* April 2012;**103**(2–3):158–63.

38. Mormede P, Desautes C, Garcia-Belenguer S, Perreau V, Sarrieau A, Moisan M, et al. Genetic influences on psychobiological responses to the environment. *Appl Anim Behav Sci* August 01, 1996;**49**(1):104–5.

39. Ohkura S, Fabre-Nys C, Broad KD, Kendrick KM. Sex hormones enhance the impact of male sensory cues on both primary and association cortical components of visual and olfactory processing pathways as well as in limbic and hypothalamic regions in female sheep. *Neuroscience* June 25, 1997;**80**(1):285–97.

40. Olmos G, Turner SP. The relationships between temperament during routine handling tasks, weight gain and facial hair whorl position in frequently handled beef cattle. *Appl Anim Behav Sci* December 15, 2008;**115**(1–2):25–36.

41. Orihuela A. Some factors affecting the behavioural manifestation of oestrus in cattle: a review. *Appl Anim Behav Sci* November 2000;**70**(1):1–16.

42. Palme R, Robia C, Messmann S, Hofer J, Möstl E. Measurement of faecal cortisol metabolites in ruminants: a non-invasive parameter of adrenocortical function. *Wien Tierärztl Monat* 1999;**86**:237–41.

43. Pedernera-Romano C, De La Torre JIR, Badiella L, Manteca X. Associations between open-field behavior and stress-induced hyperthermia in two breeds of sheep. *Anim Welfare* August 2011;**20**(3):339–46.

44. Peeters M, Verwilghen D, Serteyn D, Vendenheede M. Relationships between young stallions' temperament and their behavioral reactions during standardized veterinary examinations. *J Vet Behav* September–October 2012;**7**(5):311–21.

45. *Performance genetics: heritability of performance*. http://performancegenetics.com/heritability-of-thoroughbreds-performance/, downloaded 4/7/14.

46. Pervin LA, John OP. *Personality: theory research*. 7th ed. New York: John Wiley & Sons; 1997. p. 609.

47. Rekwot PI, Ogwu D, Oyedipe EO, Sekoni VO. The role of pheromones and biostimulation in animal reproduction. *Anim Reprod Sci* March 30, 2001;**65**(3–4):157–70.

48. Rodgers AB, Morgan CP, Bronson SL, Revello S, Bale TL. Paternal stress exposure alters sperm microRNA content and reprograms offspring HPA stress axis regulation. *J Neurosci* May 22, 2013;**33**(21):9003–12.

49. Rushen J, Taylor AA, de Passillé AM. Domestic animals' fear of humans and its effect on their welfare. *Appl Anim Behav Sci* December 1999;**65**(3):285–303.

50. Sapolsky R. *Why zebras don't get ulcers*. 3rd ed. New York: Holt Paperbacks; 2004. p. 560.

51. Scott JP, Fuller JL. *Dog behavior: the genetic basis.* Chicago: University of Chicago Press; 1965. p. 468.
52. Seong K-H, Li D, Shimizu H, Nakamura R, Ishii S. Inheritance of stress-induced, ATF-2-dependent epigenetic change. *Cell* June 24, 2011;**145**(7):1049–61.
53. Tolley EA, Notter DR, Marlowe TJ. Heritability and repeatability of speed for 2- and 3-year-old standardbred racehorses. *J Anim Sci* June 1983;**56**(6):1294–305.
54. Turner SP, D'Eath RB, Roehe R, Lawrence AB. Selection against aggressiveness in pigs at re-grouping: practical application and implications for long-term behavioural patterns. *Anim Welfare* May 2010;**19**(Suppl. 1):123–32.
55. Valenchon M, Lévy F, Neveux C, Lansade L. Horses under an enrichment program showed better welfare, stronger relationships with humans and less fear. *J Vet Behav* November/December 2012;**7**(6):e16.
56. Visser EK, van Reenen CG, Hopster H, Schilder MBH, Knaap JH, Barneveld A, et al. Quantifying aspects of young horses' temperament: consistency of behavioural variables. *Appl Anim Behav Sci* December 06, 2001;**74**(4):241–58.
57. Voisinet BD, Grandin T, O'Connor SF, Tatum JD, Deesing MJ. *Bos indicus*-cross feedlot cattle with excitable temperaments have tougher meat and a higher incidence of borderline dark cutters. *Meat Sci* August 1997;**46**(4):367–77.
58. Wolff A, Hausberger M, Le Scolan N. Experimental tests to assess emotionality in horses. *Behav Processes* September 1997;**40**(3):209–21.

Chapter 4

Learning in Livestock

If you keep doing what you've always done, you'll get what you've always gotten.

Tony Robbins

The horse's name was "Chipper," and she was an eight-year-old mare that had been a great show horse for a girl who was going off to college. Chipper had been hauled a "billion" miles over the years, but it was time for her to give another youth a chance to experience the joys of having a great horse to ride. She was going to a new home in Indiana, but getting her there was going to be a two-step process. Because the original owner was heavily involved in showing, he was going to take Chipper from Minnesota to Iowa and leave her with Frank, another horseman, until the new owners could get her a day or so later. They arrived with a brand new trailer and shiny new halter, rope, and leg wraps. At the time, Frank was gone, so the new owners got Chipper prepped and ready to load. But she would not go in the trailer. No amount of food bribery, pulling, pushing, or pleading worked—the mare stood fast. After a couple hours of this, Frank got home and went to help. He threw the lead rope over the mare's neck and in a firm voice said, "Chipper, *^@#, get in!" And Chipper calmly stepped right into the trailer. She knew how, but she also knew how to spot an inexperienced handler.

Efficient livestock handling involves the timely addition, maintenance, and withdrawal of meaningful stimuli based on the principles of learning. However, by watching livestock handlers, it is also evident that the transfer of the science of learning from academia to industry has been slow to happen.[47] Conflict behaviors can arise when livestock handlers misunderstand the appropriate application of these principles of learning and when there is a lack of consistency by and between handlers. This reduces efficiency in addition to creating welfare issues for the animals.

Discussions of learning should typically begin with the definition of what exactly is meant by the term. Learning is defined as the process of gaining knowledge or skill by studying, practicing, being taught, or experiencing something.[48] In livestock, this will take the form of a relatively permanent change in the probability of a response occurring as the result of an experience, reflecting an observable response rather than a cognitive one.[46] The process of learning in animals has been studied for many years, and the basic concepts that apply to humans also apply to livestock.[31] Differences in sensory perception, previous experiences, and genetic influences do

Efficient Livestock Handling. http://dx.doi.org/10.1016/B978-0-12-418670-5.00004-4
59

result in variations between species, but the livestock species, as animals of prey, will have more similarities than differences. Although there are variations in how the types of learning are categorized, what is most important to know are the different types of learning and how to use them in humane handling (Table 4.1).

TABLE 4.1 Comparison of Different Types of Learning

Type of Learning	Description
Associative	
Classical conditioning	Couples a stimulus and nonrelated cue with an involuntary response until the nonrelated cue alone causes the response (also called Pavlovian conditioning)
Contingency	Closeness in time of stimulus and nonrelated cue is related to length of time for connecting the nonrelated cue to the response
Operant conditioning	Voluntary activity associated with a consequence (reinforcer or punisher)
Trial and error	First outcome is by chance and the result increases the probability of the behavior happening again
Imprint	Animal learns to identify what species it belongs to, limited to specific sensitive period
Latent	Learning is not immediately obvious and the consequence is internal
Observational	Watching another successfully do a task for a reinforcer occurs faster than by trial and error
Chaining	Tiny steps toward a behavior are reinforced. Shaping is a subset where reinforcement of earlier learned behaviors is eliminated in favor of correct next tiny step.
Generalization	Response to new stimulus is the same as that to a previously learned stimulus because of the similarity of the stimuli
Non-associative	Response to a stimulus changes in strength due to the repetition of the stimulus
Habituation	Reduction in strength of response to a stimulus due to repetition
Extinction	Response to cue is reduced in strength over time as no reinforcement occurs
Sensitization	Increased strength of response to a stimulus due to repetition
Desensitization	Gradual reduction in response by gradual increase in stimulus strength
Complex (insight)	Equated with thinking

ASSOCIATIVE LEARNING

The first broad category of learning is associative learning. This is the type of learning where the response is dependent on receiving a specific cue. This may or may not be related to some type of reward or punishment, as will be more thoroughly discussed under the two major subdivisions of this type of learning.

Classical Conditioning

One of the first methods of learning described scientifically was classical conditioning, also called Pavlovian conditioning after Ivan Pavlov who first described it. Here an unconditioned stimulus results in an involuntary response, usually controlled by the nervous system. Pavlov first observed this in laboratory dogs. When food was presented (unconditioned stimulus), the dogs would drool (response). Then he started ringing a bell (a neutral stimulus which has no normal relationship to the response) at the same time food was presented. Initially when he did this, the dogs would drool because of the association of food. Eventually the dogs made the connection between the bell and food, so they drooled even if the bell was the only stimulus. There are many real-world examples of classical conditioning because it can happen unintentionally when the animal pairs two stimuli in its mind (Figure 4.1). The pressure on an udder's teats from a suckling young or milking machine (unconditioned stimulus) releases the hormone oxytocin which results in milk letdown (response). If a neutral stimulus that has nothing to do with oxytocin release (walking into

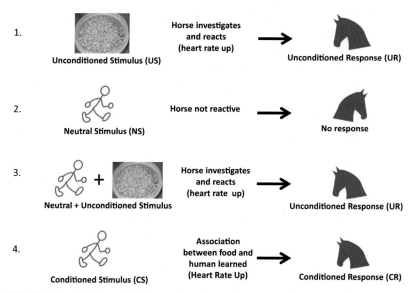

FIGURE 4.1 Classical conditioning in a horse by associating humans with a highly palatable food. © *Don Höglund (2015).*

the milking parlor) occurs about the same time as the teat pressure, the cow will eventually associate the entering of the building (conditioned stimulus) with milk letdown, and milk flow begins even before the milking machine is applied. An animal that learns that a zap from an electric fence will hurt may respond to the click of the fence charger with an increased heart rate if it makes the connection between the sound and sting. Saliva begins to flow when noise begins in the feed rooms because the horse made the connection between the grain meal and the noise.

By using an expanded definition that includes the connection of a neutral stimulus to any behavioral response, it is easy to see how classical conditioning is involved in learning. The classical example is when a dog learns to sit on command. The actual sitting posture is shaped by raising the head up and back with a food treat (a type of operant conditioning which will be discussed later) and then adding the neutral stimulus, the word "sit." Eventually the dog will sit for the command, even though no treat is given. Show horses learn to use the announcer's words to change gaits in the same way.

Classical conditioning can also be used to create a bridging stimulus, the concept of clicker training. The animal comes to associate a click with food. Gradually, the interval between the click and the food is lengthened, and the sound of the click becomes a type of reward that bridges the time from the response to the real reward. It gives the message that the response was correct and that the positive reinforcer is coming soon. The advantages of the bridging stimulus are that it can be given closer to the appropriate time for reinforcement, it is more consistent in timing and tone, it provides a little extra time to get the physical reinforcer to the animal, and it facilitates learning.[44]

Operant Conditioning

The second type of associative learning is operant conditioning. There is an expectation that two events are connected because of the intimate association between them. "If I do X, I get Y." The animal manipulates its environment, and a positive or negative outcome occurs. In other words, operant conditioning in all its various forms is all about the consequences of an action.

The Consequences

Before discussing the types of operant conditioning, it is important to understand what is meant by the *consequences*. We tend to think in terms of rewards and punishment as the types of things that can be used to encourage or discourage a behavior, but it actually gets more complicated than that. There is reinforcement and punishment, and both have positive (something is added) and negative (something is taken away) forms (Table 4.2) (Figure 4.2).

Positive reinforcement adds something and is typically equated with a food treat, praise, or gentle stroking. Actually a "reward" can be internal too, such as in a feeling of accomplishment. This internal type of positive reinforcement is

TABLE 4.2 Outcomes for an Action

Type of Outcome	Positive to the Animal	Negative to the Animal	Description of the Outcome
Positive reinforcement	+		Something positive is added (reward)—action is likely to occur again
Negative reinforcement	+		Something negative is removed—action is likely to occur again
Positive punishment		–	Something negative is added—action is less likely to occur again
Negative punishment		–	Something positive is removed—action is less likely to occur again
Nothing			Behavior is eventually extinguished unless it is internally rewarding

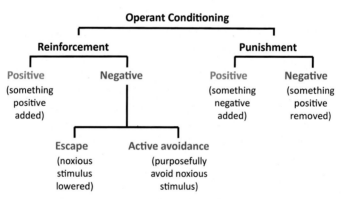

FIGURE 4.2   The relationships between reinforcement and punishment in operant conditioning. © *Don Höglund (2015).*

very rewarding, making it difficult to overpower or eliminate. As an example, pigs that are handled and brushed several times a week seek physical contact with humans more often than nonhandled pigs, even though both handled and nonhandled pigs are repeatedly chased and caught.[60] The positive interactions are internalized and reinforce the contact behavior.

Negative reinforcement removes something negative—a relief from discomfort or distress. Negative in this case does not imply bad, just a lessening of a discomfort. A rider removes leg pressure when a horse begins to move forward, as an example. The negative feel of the leg pressure goes away when the animal begins to move. A dairyman approaches a cow and she moves away because of the encroachment into her personal space. The discomfort is relieved. Negative reinforcement is a very common part of working with herd/flock species, even though its application may be very subtle and is probably unrecognized by the humans involved.[47] Even the pain of an electric prod stops when the cow moves—negative reinforcement that is not advocated.

Positive punishment adds a negative result to a behavior—what is typically just called "punishment." A horse pulls its foot away from a farrier, and the farrier hits the animal on the belly with a rasp. The hitting with the rasp is the direct application of positive punishment, so that the horse connects the behavior with the person doing the punishing. An animal can also associate a behavior's outcome with an action instead of with a person. As an example, a person who is showing a halter horse puts a thumb tack in her glove so that when the horse tries to bite, it encounters the sharp point. In this example, the horse might expect the person to hit them (a direct application of positive punishment), but instead the person does not move, and the horse associates the act of biting with being stuck by a sharp point (remote positive punishment) and not with who is holding the lead shank. A curious calf sticks its nose out to investigate an electric fence and gets zapped. The behavior is punished and the calf connects the fence with bad outcomes.

Negative punishment is when something good is removed. A person is about to put feed into a horse's stall and the horse flattens its ears. The person applies negative punishment if he stops the process and walks away before any feed leaves the bucket.

With four reinforcement/punishment options, which will be the most effective? In many species, positive reinforcement provides the best long-term result. This is especially true when rehabilitating problem horses and during encounters with frightening objects.[18,24] It is important, though, that the reinforcement must reward the desired calm behavior and not the signs of reactivity. At other times, the type of lesson is important in determining the best type of reinforcement or punishment to use, or both positive and negative reinforcements work equally well. Finally, in some situations, the outcomes are more dependent on the preferred learning styles of the individual animals.[14]

To the animal, negative experiences can be equated with positive punishments and can result in a long-term memory of that experience.[35] A mild-mannered horse may suddenly become difficult to handle, trying to escape, whenever it sees a syringe and needle because of a previous experience with a painful injection. Cattle become reluctant to enter a chute because of past experiences where electric prods were used.

There are a lot of reasons both reinforcement and punishment can fail, but inappropriately applied punishments appear to have the worst outcomes.[2,21] Punishments in particular need to be of the proper intensity for the individual.

Not severe enough means the animal learns to ignore it, even if its intensity is gradually increased. Too severe means the animal learns to avoid the punisher instead. People are also notoriously inconsistent in the application of punishments and in order to be effective, every occurrence of a behavior needs to be punished and with equal intensity.

The biggest error made in reinforcing or punishing a behavior is in the timing of its application. The direct connection between event and outcome occurs in less than one-half a second! The longer the interval between happening and consequence, the less likely it is that the animal makes the connection. A delay of only 10 s has been shown to significantly increase the degree of difficulty of a task for horses. This unpredictability of the stimulus–response relationship used in training will result in conflict behaviors and actually block learning.[50] Foals being taught to lead showed faster learning when the pull on the lead was stopped (negative reinforcement) as soon as the foal took its first step; although a longer delay worked too when specifically associated with two steps before the pull stopped.[67]

Trial and Error Learning

Now that the consequences of operant conditioning have been discussed, it is appropriate to mention the five types of learning that fall under this category. The first of these is trial and error learning. Here the first occurrence is a natural, spontaneous, random act, and that act is reinforced. A horse fiddles with a gate latch and accidently gets it open. The escape behavior is the reward. As a result of the positive reinforcement, the likelihood that the animal will try the behavior again increases, and each success increases the frequency of a reoccurrence. Trial and error learning is common in animals. To move a herd or flock, a person moves into the animals' flight distance and they start to move away (negative reinforcement). The person stops and even backs up a little to more strongly reinforce the lesson of moving away as a person approaches. The next slight movement toward the cattle also results in them moving forward. Soon slight movements can be used to guide the herd in the desired direction of the milking parlor, and because the human's movement is slow, the cattle walk instead of run. The cattle now anticipate movement when a person appears. When a heifer smells feed on the opposite side of the headlocks, she may stick her head through the gap and be rewarded for doing so with something to eat. There is now an increased likelihood that she will try again. Each success reinforces the behavior and the animal learns to associate food through the lockup.

Imprint Learning

Imprinting is the learning of what species an animal is to identify with. It occurs during a restricted time period (sensitive period) that varies somewhat with the species involved. The newborn young of livestock species are relatively well developed at birth, which was necessary for their ancestors to survive in the wild. This means that the young had to quickly learn who "mom" was and what the other animals around it looked like. These are important lessons for survival and future mating—really important to the animal's understanding of "self." Imprinting to one's own species has a high priority and can only be restricted by complete isolation from their own kind. More than one calf has been raised alone in a pen by the house and come to relate better to the dogs than to other cattle.

A great deal has been made about imprint learning in foals. Specific guidelines have been described where a person exposes the foal to a number of different stimuli, such as the handling of its feet; rubbing it with a plastic bag; touching the inside of the mouth, ears, and nostrils; and spraying it with a liquid from a sprayer bottle. This is not imprinting. It is socializing the animal to a different species (human) and habituating it to noxious stimuli (a type of non-associative learning that will be described later).[23] The imprinting occurs to the mare. While many horse farms like the results of this early handling, research has shown that without intermittent reinforcement, the effects are not lasting.[23,8,42] There is evidence that handling for as little as one hour postbirth can negatively impact foals and can result in a stronger dependence on the mare, reduced play, and impaired social competences at all ages.[20]

Imprinting has also been studied in calves to find the significance of early handling with later responses to people. Calves that are fed and touched by people during the first few days of life without their dam being present are significantly more likely to stand still and face a stranger who approaches it later.[27]

Latent Learning

Latent, or innate, learning is that which occurs but is not immediately obvious and for which there is no obvious reward. In many cases, the reward is actually an internal one—the reduction of stress. There are several excellent examples of latent learning in people who grew up around animals. They watched what different animals did and now "instinctively" know what to expect because they had actually learned what to expect in a particular situation over time. A cat naps on a person's lap and then suddenly bites the person and jumps off. As soon as it is on the ground it begins grooming itself. What does grooming have to do with biting? The behavior is called a "displacement activity" because it is out of context. People who have been around cats have probably experienced the behavior so they "knew" right away what happened—latently learned. They might not have known "displacement activity" but they could quickly connect what the word meant with the behavior.

In animals, the lessons associated with latent learning are more likely to be associated with their environments. A young animal encounters objects in its normal environment and learns not to be afraid of feed tubs, salt licks, or three wheelers because it sees these things frequently, and none of the other animals around it are reacting. The confidence gained by exposure to novel things in the environment allows the animal to be less reactive in new environments too. Horses pastured outdoors have been shown to complete trials and training programs more quickly than horses that are individually stalled.[55,59] If the animal is only exposed to loud, fast-moving people, they come to shun all people because they did not like what they were first exposed to. This is particularly important for cattle because of their sound sensitivity. There are no "do overs" in latent learning.

Observational Learning

Observational, or social, learning is more difficult to prove in animals because they rarely mimic back exactly what they see. In observational learning, an animal should learn a task faster by watching another animal successfully perform it first. Evidence that this has occurred is then suggested when the number of trials the observer animal makes is less than that required by trial and error alone.

Owners often make the assumption that when individuals within a group show the same behavior, particularly cross-suckling in calves or stereotypes such as cribbing in horses, that it is because they learned it by observing the behavior of others.[14] Well-controlled tests for observational learning in multiple livestock species have failed to positively demonstrate that it occurs. That said, studies in horses generally support the conclusion that horses do not learn by observation,[1] but there are suggestions that this type of learning might occur in horses under unique circumstances. Young colts may learn some sexual behaviors by observing older breeding stallions.[50] Foals that were present when their mothers were brushed for 15 min a day would approach, allow stroking, and accept a saddle pad placed on their back more readily than control foals.[23] Horses also pay attention to the actions of dominant horses in their herd significantly more often than they pay attention to lower ranking ones.[29] They will choose a feed bucket that the dominant horse did not eat from and avoid locations where a high ranking horse typically eats. Whether this is a result of observations or something else, such as odors, is uncertain.

It has not been determined if observational learning really occurs in other species, but it has been suggested in ruminants and pigs.[7,53] Observations might help young bulls breed successfully faster,[41] though other reports suggest this type of learning does not occur in cattle.[65]

Chaining

The fifth type of operant conditioning is *chaining* (also called shaping). It is a technique that typically uses the learning of one behavior as a step to learning the next, linking them into a series that has the desired outcome. B.F. Skinner taught chickens to peck a specific location when a light was turned on by first rewarding the smallest head move toward the pecking spot and then requiring that the head come even closer to the desired location for each subsequent reward. Eventually the body also had to face the goal too. This procedure uses trial and error learning to progress in a stepwise fashion toward a desired goal. The technique can be very helpful in teaching animals to do various tricks.[14] Chaining a complex behavior involves the phenomenon of "learning to learn." As the lessons of each step build on the one before it, each step becomes easier and is accomplished in a shorter time.[14] If, however, previous learning is not consistent with the new lesson to be learned, it can actually interfere with new learning.[57] That is also the primary reason that older animals may take longer to learn something compared to their younger herd mates.[45]

Generalization

Generalization is the third type of associative learning. It is represented by the occurrence of a behavior in a new location or situation because of the similarity of stimuli. Mary taught the dog to "sit" in the kitchen before she put down its morning meal. John tells the same dog to "sit" by a gate as he opens it and the dog responds. While the word "sit" sounds the same to us, to the dog the tone quality, speed of delivery, loudness, and other tone qualities are varied. The similarity of the pronunciation was generalized between the situations. Pigs learn to associate one color of coveralls with gentle handling and ultimately interact more readily with any person wearing the same colored coveralls, even in different locations.[25] A rider practices crossing a wooden bridge for a trail class but at the horse show, the animal refuses a similar obstacle. The refusal occurs because the horse has not generalized that a raised wooden obstacle is similar, regardless of its location.

NON-ASSOCIATIVE LEARNING

The common theme for types of non-associative learning is that the response to a specific stimulus changes over time due to the repetition of that stimulus. The behavior could either increase or diminish in intensity. The types mentioned here are the most common, at least relative to the livestock species. For each type of non-associative learning to be successfully used, it is extremely important to not stop their application too soon. To do so will actually reinforce the very thing that is undesirable.

Habituation

A stimulus is repeated and repeated until there is no longer a response. This is the basis for habituation learning. The first time a noise startles an animal, it jumps and focuses its attention toward the source. The second time it happens in relative rapid succession, the jump is less dramatic. Eventually, the sound is ignored altogether. This is a natural learning process that helps an animal subconsciously filter out normal, inconsequential events in its environment so it can focus on potentially dangerous ones. It also has application in domestic animals. At a horse show, there are plastic streamers lining the refreshment stand next to the arena, and they are flapping in the breeze. The initial response of a horse going by is to shy. If walked back and forth past the streamers, the horse habituates to the flutter and comes to ignore it. In another example, a horse that avoids walking over a plastic tarp in a trail class can be habituated to doing so if it is kept in a stall in which a plastic tarp covers the floor. Each step results in the crinkling noise, and eventually the horse quits paying attention. Studies in horses show that yearlings that were raised in isolation from other horses react less and habituate faster to novel objects than group-raised yearlings. They are also better learners, perhaps because they show less emotional reactivity.[33] Studies also suggest that the color of the novel object may be of more significance in habituation learning than the shape of the object.[4]

Extinction

A specific cue is used to get a response, and the response is reinforced with some high value reward. Once the response is well learned, the owner stops all reinforcement. Without the reward, the animal eventually stops showing the behavior for the cue. The behavior has been extinguished. This extinction can work in a negative or positive way depending on what the behavior is. As a negative, the dog that learned to "sit" on command for a food treat will stop responding to the word if some reward is not occasionally given. Undesirable behaviors sometimes have a reward that the owner does not realize. A horse has learned to open a stall door by wiggling the latch with its mouth through trial and error. The latch is lowered and a snap put on it so that the horse can no longer undo it. Initially, the horse might try harder to open the latch, but eventually extinction occurs and the horse quits trying because of the lack of success over time.

Sensitization

Sensitization is the increased strength of a response as the result of repeated exposure to a specific stimulus. If a person pats a goat on the head, eventually the goat is sensitized enough that it butts the person. The next time, a few pats are necessary to result in the butting. Eventually the mere sight of a person will trigger the rearing-head butting behavior because the goat has been sensitized to its head being touched.

Desensitization

Desensitization is commonly used by behaviorists to treat undesired behaviors. In a controlled setting, a stimulus is presented in such a way that the usual reaction does not occur. If a noise is a problem, the volume is started so low that there is no reaction. The intensity of the stimulus is very gradually increased over a long period of time in such a way that it is ignored and the behavior is no longer triggered. It was previously pointed out that punishments may not be effective because they were either not severe enough or too severe. If a positive punishment is used too lightly, the animal ignores it, so the person uses it a little harder. The animal ignores it. Increasing levels of harshness will reach a level much more severe than would have been necessary originally because the animal is desensitized to the punishment.

Desensitization is similar to habituation in that the stimulus is repeated over and over again, but in desensitization, the strength of the stimulus increases gradually whereas in habituation, the strength does not change. Using the horse that is afraid to walk on plastic described above as an example, if the stall floor is covered only with a tarp, habituation occurs. If, however, the tarp is covered with shavings and the amount of shavings used is gradually reduced over several days, the technique would be desensitization.

COMPLEX LEARNING

The third major type of learning is complex, or insight, learning. This is usually equated with thinking. Obviously it is extremely difficult to prove in animals because we cannot ask them directly, so the presence or absence of thinking is inferred by behaviors. As an example, in video recordings of horses working in a "T" maze, during one trial a single horse was seen stopping after entering the stem of the "T," looking left and then right, and finally proceeding toward the appropriate goal. A horse would typically enter and randomly move into one branch of the "T" until it learned to go to the side that contained a specific black or white bucket containing food. Does this prove complex learning? No, but it could be. Chimpanzees modifying grass and twigs to successfully fish for termites were first reported by Jane Goodall many years ago. This discovery of tool use was the beginning of scientists questioning whether animals were actually capable of complex thoughts and how they could prove or disprove it. People came to accept the possibility that greater apes might be capable of complex learning, but could lower animals be capable of it as well? Watching other animals suggests that thinking is not restricted to animal species closest to humans on the phylogenetic scale. Dogs have moved wood from the center of a yard to the fence line so that they could get over a fence. Birds have dropped rocks into jars of water to float worms high enough to reach them. Pigs can learn to see hidden food in a mirror and go directly to the food and not to the mirror.[7] Evidence also exists that suggests that the rearing environment affects complex learning, in that some enrichment with novel objects will cause animals to be generally more active and calmer when handled.[63]

Concept learning trials also study when an animal first learns to distinguish between two different things, such as a solid colored circle and an open circle, and then transfers that distinction to other objects, such as a solid colored triangle and an open triangle. Studies in horses show they have some ability to make these distinctions, but not in all situations.[14,6,11,12,52] Animals often do not have the opportunity for exposure to a variety of situations that give them the background necessary to develop concept learning. Horses that learn to load into a single horse trailer typically balk at entering strange trailers. Those that have been in several different trailers are comfortable going into other trailers, as well as into other narrow areas.

Social learning might be placed in the complex learning category as well. While some of the things an individual may learn can happen through observational or trial and error learning, group membership lessons would include things like place within the social hierarchy, individual recognition within and across species, and spatial features of the habitat.[59,9,34] Interactions like mutual grooming and play are associated with endorphin release in the brain, which is an internal reward mechanism. If these things are blocked, the buildup of stress is shown as a rebound of excessive behaviors when the social interactions are again allowed.[64] This suggests an internal reward. Horses that are raised with other horses learn tasks quicker than those housed individually.[29] Although this has been shown in other species too, the significance of social learning is probably underestimated in livestock. Those animals that watch higher ranking horses follow a human will copy the following behavior, while those that watch lower ranking horses follow do not.[30] In bucket selection tests where people hold the buckets, the horse is more likely to go to the location of any person looking at them or to a familiar person who is not even looking at them rather than pay attention to specific bucket selection.[28]

OTHER FACTORS

In order for learning to occur, there are a number of things than must come together. There generally needs to be a conscious recognition on the part of the animal that there is something going on. This, in turn, requires some type of sensory input. Although learning can be active (with motivation) or passive, learning and memory under passive conditions are poor. It has been suggested that keeping the brain actively involved in the learning tasks through variations in activities and environments, in addition to the appropriate use of reinforcers and punishers that encourage active learning, is important.[50] Learning studies often fail to take into account the physiological uniqueness of each species. For example, colored choices have long been a problem. Another example would be the use of a light cue placed high on a wall over a "chosen" side in a T maze that is turned on to indicate the correct choice. Because it is high up, it becomes irrelevant to species with a horizontal pupil, and the results of the experiment must be considered flawed.[22] The duration of the teaching sessions is also relative to the ease of learning. A moderate amount of handling also seems preferential to extensive amounts, at least in yearling horses.[17] Longer sessions are associated with decreased learning efficiency.[45]

Age affects learning in many species. Young animals spend more time investigating things and learn quicker.[17,40] This is because they have fewer previous experiences within the environment with which to compare new things or compete with new lessons. In cattle, 15-month-old heifers learn a task quicker than cows after their second calving. On the other hand, the cows outperformed the heifers in long-term memory tests.[26]

Laterality also impacts learning. It has been shown that goats will go more easily to the left branch of a "Y" maze than to the right side. This is consistent with the right-brain dominance for novel situations; however, once the lesson is learned, teaching the goats to go to the opposite branch of the "Y" (reversal learning) was the same, regardless of which branch they were originally trained to use.[32]

Other, less defined things can affect learning. Estrus disrupts social orders and hormone fluctuations can even alter sensory perceptions and learning.[16] External temperature extremes, the degree of hunger or thirst, and potential threats like the close proximity of a high ranking herd member are other factors that should be considered.

How long the various lessons are remembered is likely variable. Sheep can remember the faces of 50 other sheep for more than two years, but a stallion may not recognize young males from his group 18 months after they left.[34] Is that because the social bonding within a flock is more significant than that in a horse herd, or is there some other reason? Highly negative things are remembered for a long time, but how long other types of memories are kept by each species has not been studied.

Domestic livestock will encounter a number of things that are contrary to their evolutionary adaptive behaviors. As examples, horses maneuver around and over objects that would not exist in the wild, and cattle walk down narrow alleys and into trailers in poor lighting. These are things that are the most easily accomplished if the animal is allowed to learn not to overreact to the situations so they can discriminate and respond to a wide variety of different stimuli.[45]

Motivation

Exposure to something new does not necessarily result in learning, even latent learning, unless there is some level of motivation. This could vary from a passing curiosity to a high level internal drive. If food is used as a reward but the animal either is not hungry or does not like the food, the lesson fails. Some pigs are very good at finding underground truffles because they really like to eat truffles. Dogs can be taught to sniff out truffles too, but they are harder to motivate to keep on task because the taste is not a high value to them. If an animal is eating, drinking, or breeding, the physiology of the animal is the driver of the behavior, but the mental motivation for such behaviors is unknowable.

Emotional State

One of the more nebulous factors affecting learning and social interactions is the individual's "emotional state" or "mood." Everyone seems to know what emotion is until they are asked to define it,[36] and not being able to define it makes it hard to study. This is especially true in nonverbal animals. For example, "fear" as a feeling has a dozen English-language synonyms. In reality, the true meaning in any unique situation can only be known by introspection.[62] Trying to teach something to a horse on a cold, windy day when little things tend to spook the horse is much harder than during a mild summer day.

As with other words used to describe various emotions, an understanding of the word "fear" is complicated because people use "feeling words" interchangeably as descriptors.[36,38] "Fear," "happiness," and "sadness" are emotions that humans can experience. While physiological responses in animals may parallel those of humans in emotionally equivalent situations, we do not actually know if the mental experiences are the same. It is more appropriate to use descriptions like "attention-seeking behavior" and "asocial behavior." "Fear" is the thought and "threat-elicited reaction" describes the response. Even in humans, fear is not likely to exist with the first experience; it builds based on that learning and then can be subsequently self-rewarding (Figure 4.3). A cow avoids the parlor entrance. Assuming that "fear" is her motivation misses the cause of the avoidance behavior, and the solution used may be inappropriate. Perhaps the precipitator of the avoidance is merely due to the parlor being unfamiliar, in which case force or pain are inappropriate. Perhaps pain or startle was involved in a previous parlor experience, giving a different reason for the behavior. Assuming "fear" is the cause misses the real motivation for the avoidance behavior in the first place. In the end, it just does not matter what the animal was thinking or feeling. What can be observed, measured, and documented is what is important.

FIGURE 4.3 The feeling associated with fear and the resulting escape behavior are unique from each other. The feeling requires introspection, while the behavior is observable. © *Don Höglund (2015).*

The goal is to identify the precipitating circumstances in the environment that result in the unwanted behavior and remedy the situation.

The confusion in using emotionally descriptive terms is that because humans obviously have feelings, the assumption is made that other animals exhibiting observable defensive responses are also feeling fear. This is not a denial of consciousness in animals, but rather an attempt to accurately depict what is occurring rather than to infer something that may or may not be.[39,37]

LEARNING IN LIVESTOCK SPECIES

While learning occurs in all livestock species, it has been most extensively studied in horses because of the types of tasks they are asked to do. There is a great deal of anecdotal information about equine learning and training, especially compared to the amount of science done on the subject.[5] Add to this the complication that training typically is based on the use of negative reinforcement strategies, while experimental designs rely almost exclusively of the use of positive reinforcement.[47,50,45] This begs the question about the long-term impact of one method over the other relative to the training of horses. The answer seems to depend on the type of lesson the horse is to learn. The association of a reward given by a person during any type of interactional context enhances learning, and it results in increased contact and interest toward the person immediately and even after eight months of separation.[56] Horses with trailer-loading problems also showed faster learning and less stress when positive reinforcement was used.[19] This suggests that positive reinforcement has a powerful impact on learning, much like it does in dogs.[2] However, relief from an aversive stimulus is also remembered for a very long time.[66]

Learning has been an important part of the survival of any species that evolved in open conditions, and horses have become good at changing their behaviors because of experience. They can learn to habituate to some stimuli and yet remain alert for others, suppress instincts, make connections between events, learn tasks that are not natural behaviors, and respond fast to reinforcers.[31,14,49] Stereotypic behaviors occur primarily in stabled horses, where there is not a lot of environmental enrichment. Besides being health and welfare concerns, the development of any stereotype has been shown to lower a horse's ability to learn.[15,43]

A common rumor spread among horse people is that what a horse learns by seeing it with one eye, it must also learn by seeing it with the other eye too. The two learning situations are considered to occur separately. This is not accurate and is not consistent with what happens in other species.[14,51] Interhemispheric transfer of visual information has been shown in horses too.[14,13,10] This makes sense anatomically because there is a decussation of about 17% of fibers from the optic nerves to the contralateral side of the brain, so information from each eye goes to both sides of the brain. There are also fibers that cross the brain's corpus callosum, which transfer visual information between the optic areas of the brain and the reacting parts on both sides of the brain.[61]

Like other species, cattle are good at remembering aversive handling, and that can create problems later. The unfortunate part of this situation is that the stockperson is often not aware of how his or her actions are being perceived. While positive interactions, such as hand feeding, grooming, and gentling, do not result in improvements above baselines, studies find that aversive handling will negatively impact the amount of time and the force required to move cattle.[54] Shouting, hitting, and electric shocks were among those behaviors found to be most aversive by cattle.[54] While hitting and shocks would be considered by most people to be aversive techniques, shouting and loud noises are often not thought of that way. However, as cattle can hear (as was described earlier), they are aversive.

Goats, like other species of livestock, are easily trained to perform complex chains of behavior and have good long-term memory. They also adapt to changes in the environment and location readily.[58] Because many of the memory tests involve the use of food, it has been speculated that goats, and probably other ungulates as well, possess these learning and memory skills based on their evolutionary need to forage efficiently and safely in harsh climates.[3]

Learning in pigs has become a research topic that has shown the breadth of this species' ability to learn. Pavlov originally did not consider pigs to be good experimental subjects, but more recent studies show classical conditioning occurs in this species just as it does in others.[7] Feral hogs are very skilled at finding ways to get through obstacles set up to block their access to feeding sites. If a fence is in the way, they have been known to create holes in woven wire to gain entrance to a prized food source (Figure 4.4).

FIGURE 4.4   Feral hogs spread the wire to gain access to acorns that had fallen from several oak trees.

REFERENCES

1. Ahrendt LP, Christensen JW, Ladewig J. The ability of horses to learn an instrumental task through social observation. *Appl Anim Behav Sci* June 2012;**139**(1–2):105–13.
2. Blackwell EJ, Twells C, Seawright A, Casey RA. the relationship between training methods and the occurrence of behavior problems, as reported by owners in a population of domestic dogs. *J Vet Behav* September–October 2008;**3**(5):207–17.
3. Briefer EF, Haque S, Baciadonna L, McElligott AG. Goats excel at learning and remembering a highly novel cognitive task. *Front Zool* 2014;**11**:20.
4. Christensen JW, Zharkikh T, Ladewig J. Do horses generalize between objects during habituation? *Appl Anim Behav Sci* December 01, 2008;**114**(3–4):509–20.
5. Cooper JJ. Equine learning behavior: common knowledge and systematic research. *Behav Processes* September 2007;**76**(1):24–6.
6. Gardner LP. The responses of horses in a discrimination problem. *J Comp Psychol* February 1937;**23**(1):13–34.
7. Gieling ET, Nordquist RE, van der Staay FJ. Assessing learning and memory in pigs. *Anim Cogn* March 2011;**14**(2):151–73.
8. Goodwin D. Equine learning behavior: what we know, what we don't and future research priorities. *Behav Processes* September 2007;**76**(1):17–9.
9. Hagen K, Broom DM. Cattle discriminate between individual familiar herd members in a learning experiment. *Appl Anim Behav Sci* June 2003;**82**(1):13–28.
10. Hanggi EB, Ingersoll JF, Waggoner TL. Color vision in horses (*Equus caballus*): deficiencies identified using a pseudoisochromatic plate test. *J Comp Psychol* 2007;**121**(1):65–72.
11. Hanggi EB. Categorization learning in horses (*Equus caballus*). *J Comp Psychol* September 1999;**113**(3):243–52.
12. Hanggi EB. Discrimination learning based on relative size concepts in horses (*Equus caballus*). *Appl Anim Behav Sci* September 26, 2003;**83**(3):201–13.
13. Hanggi EB. Interocular transfer of learning in horses (*Equus caballus*). *J Equine Vet Sci* August 1999;**19**(8):518–24.
14. Hanggi EB. The thinking horse: cognition and perception reviewed. In: *Proceedings of the 51st Annual Convention of the American Association of Equine Practitioners,* Seattle, Washington. December 2005. p. 246–255.
15. Hausberger M, Gautier E, Müller C, Jego P. Lower learning abilities in stereotypic horses. *Appl Anim Behav Sci* November 2007;**107**(3–4):299–306.
16. Hedberg Y, Dalin A-M, Öhagen P, Holm KR, Kindahl H. Effect of oestrous-cycle stage on the response of mares in a novel object test and isolation test. *Reprod Domest Anim* October 01, 2005;**40**(5):480–8.
17. Heird JC, Lennon AM, Bell RW. Effects of early experience on the learning ability of yearling horses. *J Anim Sci* 1982;**53**(5):1204–9.
18. Heleski CR, Bello NM. Evaluating memory of a learning theory experiment one year later in horses. *J Vet Behav* July–August 2010;**5**(4):213.
19. Hendriksen P, Elmgreen K, Ladewig J. Trailer-loading of horses: Is there a difference between positive and negative reinforcement concerning effectiveness and stress-related signs? *J Vet Behav* September–October 2011;**6**(5):261–6.
20. Henry S, Richard-Yris M-A, Tordjam S, Hausberger M. Neonatal handling affects durably bonding and social development. *PLoS One* 2009;**4**(4):e5216. http://dx.doi.org/10.1371/journal.pone.0005216.

21. Hockenhull J, Creighton E. Training horses: positive reinforcement, positive punishment, and ridden behavior problems. *J Vet Behav* July–August 2013;**8**(4):245–52.

22. Hothersall B, Nicol C. Equine learning behavior: accounting for ecological constraints and relationships with humans in experimental design. *Behav Processes* September 2007; **76**(1):45–8.

23. Houpt KA. Imprinting training and conditioned taste aversion. *Behav Processes* September 2007;**76**(1):14–6.

24. Innes L, McBride S. Negative versus positive reinforcement: an evaluation of training strategies for rehabilitated horses. *Appl Anim Behav Sci* August 2008;**112**(3–4):357–68.

25. Koba Y, Tanida H. How do miniature pigs discriminate between people? The effect of exchanging cues between a non-handler and their familiar handler on discrimination. *Appl Anim Behav Sci* January 18, 1999;**61**(3):239–52.

26. Kovalčik K, Kovalčik M. Learning ability and memory testing in cattle of different ages. *Appl Anim Behav Sci* April 1986;**15**(1):27–9.

27. Krohn CC, Boivin X, Jago JG. The presence of the dam during handling prevents the socialization of young calves to humans. *Appl Anim Behav Sci* March 01, 2003; **80**(4):263–75.

28. Krueger K, Flauger B, Farmer K, Maros K. Horses (*Equus caballus*) use human local enhancement cues and adjust to human attention. *Anim Cogn* March 2011;**14**(2):187–201.

29. Krueger K, Flauger B. Social learning in horses from a novel perspective. *Behav Processes* September 2007;**76**(1):37–9.

30. Krueger K, Heinze J. Horse sense: social status of horses (*Equus caballus*) affects their likelihood of copying other horses' behavior. *Anim Cogn* July 2008;**11**(3):431–9.

31. Ladewig J. Clever Hans is still whinnying with us. *Behav Processes* September 2007; **76**(1):20–1.

32. Langbein J. Investigations on training, recall and reversal learning of a Y-maze by dwarf goats (*Capra hircus*): the impact of lateralization. *Behav Processes* March 2012; **89**(3):304–10.

33. Lansade L, Neveux C, Levy F. A few days of social separation affects yearling horses' response to emotional reactivity tests and enhances learning performance. *Behav Processes* September 2012;**91**(1):94–102.

34. Leblanc M-A, Duncan P. Can studies of cognitive abilities and of life in the wild really help us to understand equine learning? *Behav Processes* September 2007;**76**(1):49–52.

35. LeDoux JE. Coming to terms with fear. *Proc Natl Acad Sci* February 25, 2014;**111**(8): 2871–8.

36. LeDoux JE. Emotion: cues from the brain. *Annu Rev Psychol* February 1995;**46**:209–35.

37. LeDoux JE. Feelings: what are they & how does the brain make them? *J Am Acad Arts Sci* Winter 2015;**144**(1):96–111.

38. LeDoux JE. Rethinking the emotional brain. *Neuron* February 23, 2012;**73**(4):653–76.

39. LeDoux JE. The slippery slope of fear. *Trends Cogn Sci* April 2013;**17**(4):155–6.

40. Lindberg AC, Kelland A, Nicol CJ. Effects of observational learning on acquisition of an operant response in horses. *Appl Anim Behav Sci* January 1999;**61**(3):187–99.

41. Mader DR, Price EO. The effects of sexual stimulation on the sexual performance of Herford bulls. *J Anim Sci* August 1984;**59**(2):294–300.

42. Mal ME, McCall CA, Cummins KA, Newland MC. Influence of preweaning handling methods on post-weaning learning ability and manageability of foals. *Appl Anim Behav Sci* June 1994;**40**(3–4):187–95.

43. Malamed R, Berger J, Bain MJ, Kass P, Spier SJ. Retrospective evaluation of crib-biting and windsucking behaviours and owner-perceived behavioural traits as risk factors for colic in horses. *Equine Vet J* November 2010;**42**(8):686–92.

44. McCall CA, Burgin SE. Equine utilization of secondary reinforcement during response extinction and acquisition. *Appl Anim Behav Sci* September 10, 2002;**78**(2–4):253–62.

45. McCall CA. A review of learning behavior in horses and its application in horse training. *J Anim Sci* January 1990;**68**(1):75–81.

46. McGreevy P. *Equine behavior: a guide for veterinarians and equine scientists*. 2nd ed. New York: Saunders; 2012. p. 378.

47. McLean AN. The positive aspects of correct negative reinforcement. *Anthrozoös* 2005;**18**(3):245–54.

48. Merriam Webster. http://www.merriam-webster.com/dictionary/learning, definition, downloaded 02.11.14.

49. Miller RM. The amazing memory of the horse. *J Equine Vet Sci* August 1995;**15**(8):340–1.

50. Murphy J, Arkins S. Equine learning behavior. *Behav Processes* September 2007;**76**(1): 1–13.

51. Murphy J, Hall C, Arkins S. What horses and humans see: a comparative review. *Int J Zool* 2009;**2009**(Article ID 721798):1–14.

52. Nicol CJ. Equine learning: progress and suggestions for future research. *Appl Anim Behav Sci* September 10, 2002;**78**(2–4):193–208.

53. Nicol CJ. The social transmission of information and behavior. *Appl Anim Behav Sci* September 1995;**44**(2–4):79–98.

54. Pajor EA, Rushen J, de Passille AMB. Aversion learning techniques to evaluate dairy cattle handling practices. *Appl Anim Behav Sci* September 2000;**69**(2):89–102.

55. Rivera E, Benjamin S, Nielsen B, Shelle J, Zanella AJ. Behavioral and physiological responses of horses to initial training: the comparison between pastured versus stalled horses. *Appl Anim Behav Sci* September 10, 2002;**78**(2–4):235–52.

56. Sankey C, Richard-Yris M-A, Leroy H, Henry S. Positive interactions lead to lasting positive memories in horses *Equus caballus*. *Anim Behav* April 2010;**79**(4):869–75.

57. Sappington BKF, McCall CA, Coleman DA, Kuhlers DL, Lishak RS. A preliminary study of the relationship between discrimination reversal learning and performance tasks in yearling and 2-year-old horses. *Appl Anim Behav Sci* June 1997;**53**(3):157–66.

58. Sherry CJ, Walters TJ, Rodney Jr GG, Henry PJ. Behavioral chaining in the goat (*Capra hircus*). *Appl Anim Behav Sci* June 1994;**40**(3–4):241–51.

59. Sigurjónsdóttir H. Equine learning behavior: the importance of evolutionary and ecological approach in research. *Behav Processes* September 2007;**76**(1):40–2.

60. Tanida H, Miura A, Tanaka T, Yoshimoto T. The role of handling in communication between humans and weanling pigs. *Appl Anim Behav Sci* June 1994;**40**(3–4):219–28.

61. Timney B, Macuda T. Vision and hearing in horses. *J Am Vet Med Assoc* May 15, 2001;**218**(10):1567–74.

62. Tinbergen N. *The study of instinct*. New York: Oxford University Press; 1951. p. 228.

63. Tönepöhl B, Appel AK, Welp S, Voß B, von borstel UK, Gauly M. Effect of marginal environmental and social enrichment during rearing on pigs' reactions to novelty, conspecifics and handling. *Appl Anim Behav Sci* September 2012;**140**(3–4):137–45.

64. VanDierendonck MC, Spruijt BM. *Appl Anim Behav Sci* May 2012;**138**(3–4):194–202.

65. Veissier I. Observational learning in cattle. *Appl Anim Behav Sci* January 1993;**35**(3): 235–43.

66. Visser EK, van Reenen CG, Schilder MBH, Barneveld A, Blokhuis HJ. Learning performances in young horses using two different learning tests. *Appl Anim Behav Sci* March 01, 2003;**80**(4):311–26.
67. Warren-Smith AK, McLean AN, Nicol HI, McGreevy PD. Variations in the timing of reinforcement as a training technique for foals (*Equus caballus*). *Anthrozoös* 2005;**18**(3):255–72.

Chapter 5

Horse Handling: Practical Applications of Science

There is a touch of divinity even in brutes, and a special halo about a horse, that should forever exempt him from indignities.

Herman Melville

Hanging between two reins is a thought, Harry Whitney said. If we can visualize that, then what follows naturally is the understanding that those reins never, ever turn the head or the horse. Rather, they simply steer the horse's thought. Once established, one can ever so gently ask the horse to think around to the right with just a finger on the rein, then watch his head turn as he thinks around that way. No bracing in the neck. No head popping up in opposition to the rein. No. You're not manipulating the flesh in any physical way. There is never a need for a more severe bit. Why? Because all the rein does is present a slight suggestion. If that horse is with you, his brain understands the lightest request, and he is ready to commit to the suggestion whole heartedly and bring his own body along, too.[16]

Entire philosophies on human and livestock interaction have relied on interpreting animal behavior in the context of human behavior.[5] There is, however, a lack of evidence that animals understand their training, are aware of human intention, or experience human motivations. This is not a denial of conscious awareness in animals, but a reason to not depend on the use of human feelings in an attempt to explain animal reactions.[8]

Near the beginning of the twentieth century, we were warned not to "humanize the brute."[17] Even the Nobel laureate Nico Tinbergen proposed limitations on anthropomorphism by stating that "hunger, like anger, fear, and so forth, is a phenomenon that can be known only by introspection. When applied to another species, it is merely a guess about the possible nature of the animal's subjective state."[19] We assume that symptoms of conscious and introspective awareness are inseparable from behavioral responses, both observable and physiological. It is now known that defensive reactions and feelings are not part of the same neuronal path in humans and are not controlled by the same brain circuits.[8] Survival of any species depends on the ability to detect and respond to threats, irrespective of the animal's ability to experience fear.

Efficient Livestock Handling. http://dx.doi.org/10.1016/B978-0-12-418670-5.00005-6

In practical terms, any handler can know what their horse is doing and perhaps what event precipitated the behavior, but it is impossible to know what a horse is feeling emotionally or thinking about as a result of merely observing the animal's behavior. Instead of trying to describe mental motivations of the horse, it is more accurate to describe its behaviors.[9,10] Understanding efficient, humane horse handling requires scientific and practical knowledge of animal behavior, including how animals learn.

The first order of horse handling evaluation should be to avoid confusing handler's athletic ability with efficient handler results. During any handling process, the absence of avoidance behaviors is a potential sign that handlers can use to gauge a "so-far, so-good" philosophy. If the behavior of the horse is as intended by the handler, it should be reinforced. If the behavior is unwanted, the handler should review what was done to get the unintended result and avoid repeating that stimulus again. The essence of good handling is in the character of the handler who is willing to assess what happened and start over if necessary. That has positive implications in the well-being of horses everywhere.

A TRUE REVOLUTION IN HORSEMANSHIP

The nearly 6000-year relationship of humans with horses has established traditional equestrian techniques.[1] Recent claims of a revolution in horsemanship coincides with an evolution in the role horses play in human life.[15] Although some of the newer horse handling methods are more humane and some are no-doubt effective, many of them have bypassed the research findings, including the fundamentals of learning.[12] Understanding how animals learn should be the first priority of animal handlers. When omitted, horses face inconsistent stimuli, which can contribute to conflict behaviors. A second priority is the communication of training techniques and how they relate to learning.

Anthropomorphisms and trendy horse training concepts tend to ignore learning concepts and depend on jargon in an attempt to legitimize the technique.[13] Assumptions that horses are willing, benevolent, or malevolent participants in search of human leadership should lead to welfare concerns for the animal.[14] Interaction in any environment involving handlers and the horse creates an opportunity for learning, and animals are learning every waking minute. Only fatigue and drug interference may hinder this.[18] The lesson that is learned can be the problem. Some learning may have intended, productive outcomes. Unfortunately, some may create real problems, particularly if pain or startle is involved.

The important concepts of learning are detailed in Chapter 4, but an emphasis of some of those concepts is also given here because of their importance in horse handling. A stimulus is any detectable change in an animal's environment, and any response is a behavior or physiological event.[13] Stimuli are important because they trigger responses made by the horse. A flapping plastic bag hung up on a barbwire fence is one. So is gentle touch-pressure that starts forward movement.

FIGURE 5.1 Horses usually try to escape stimuli that startle them. The instinct to flee has evolved in fleet-footed species that are animals of prey. © *Don Höglund (2006).*

Timing of a stimulus, consistency of technique, and immediate reinforcement or punishment are critical elements in adaptive and maladaptive horse conditioning. Negative reinforcement—the reduction or removal of a negative stimulus in response to a behavioral response—is commonly used when working with horses.[3,12,14] If a human steps toward an untamed horse and the horse moves away to relieve the encroachment, the horse is rewarded through negative reinforcement. This reinforces the behavior that preceded it.

Though it is a critical component of horse behavior management, the subtleties of negative reinforcement in horses are very poorly understood.[13] Startle reactions are reflexive, involuntary, and typically result from novel, sudden, or intense antecedent stimuli. The involuntary physiological response lasts longer than the initial behavioral one. It may last 20 min or the rest of the horse's life.[7] The behavioral response of a horse is usually escape, avoidance, or attempted escape (Figure 5.1). Aversive techniques, such as striking a horse on the head, poking it in the eye, chasing it with the lead line, abusing a lip chain and twitch, or striking a horse in the abdomen, can be associated with attempted escape behavior. Even worse, they create an animal that is prone to startle or escape.

THE SENSES AND THE REACTIONS

Understanding horse behavior requires recognizing the role equine senses play in horse perception.[13] Irrespective of what livestock species is being discussed, what handler action is intended, or what animal behavior is occurring, the senses are needed to bring information to the brain to determine if it is familiar, novel, or threatening for a decision about how to react. In other words, senses are the animal's peripheral filters. The senses of vision and hearing are arguably the two most important ones involved in learning the subtle differences between safe and dangerous environmental events to enable a horse's survival.[2]

It is important to understand that as an animal that evolved with escape as a primary survival behavior, horses developed sensory systems that are different from those of humans.

The reactive distances that influence a horse's response to an approaching object are governed by this sensory input. Escape behavior is the most common response of a horse to a threat, including to a startle or painful stimulus. It can also be learned by experience. When escape is not possible, the horse is likely to exhibit aggression. Safe handling of horses with an unknown reaction history can be checked with a quick test. Limit environmental stimuli and merely approach the point of the horse's shoulder from a safe distance. This is usually safer than a head approach. The moment the animal clearly reacts to the handler's approach with a head-lift, eye and ear changes, or body movement, it demarks the horse's perceptive distance. The distance at which the animal attempts to move away is the edge of the flight distance. In order to physically work on the horse, the flight distance needs to be reduced to zero. This is done slowly with slight encroachment and withdrawal being repeated. As the animal becomes familiar with minimally reactive handlers, escape tendencies tend to fade. Handlers that use slow, predictable movements tend to have more success in approaching reactive animals. After all, handler motion can be an associated sign to what happens next.

EFFECTIVE HANDLING

The primary objective of effective handling is to get the job done irrespective of any unintended consequences, but do not confuse effective with efficient. Young horses raised in a restricted space, such as a stall, have a tendency to run everywhere when given access to a relatively large area. Handlers turn the horse loose in a round pen and then make it circle or even chase the horse with the intention to burn off some of its unspent energy. This teaches the horse to run every time it encounters a human. It works for the horse: it is rewarding because they get away. Teaching horses to run gives the appearance that the animal is "happy" to be chased, but when a prey animal is chased, stress levels are likely to go up. The chase may have been effective in providing exercise, just not in keeping the animal calm. Effective methods for handling animals are found everywhere there are horses and have been for thousands of years. They may get the intended job done, but they might coincidently be teaching an undesired lesson too. Descriptions of effective techniques are often anthropomorphic and circuitous.[4] As an example, "the horse is fearful and her fear causes her to pull away, and the reason we know she is fearful is because she is pulling away." Instead of trying to define what the horse is thinking or feeling, because we truly will never know, it is better to define the specific behavior. Then we can try to figure out what triggers it. In the previous example, fear is not the cause of the horse pulling away but an aspect of its response to events in the environment. As a second illustration, a horse is observed trying to escape from the entrance of the stall. Though she is incapable of telling us why she is struggling to escape from the entrance, a horse turning away from the stall entrance could be as simple as a naturally occurring avoidance behavior to the new environment. It could also be a learned avoidance behavior resulting from a previous negative experience in the stall, such as a phantom voltage electrifying a feeder. Or, the escape behavior could be a learned behavior precipitated by some prior trailer loading event where the horse had attempted to escape, was restrained, and struggled. In this case, the struggle is associated with being restrained, and the horse learned to avoid entrances that had a header at the top of the opening.

EFFICIENT HANDLING

Efficient handling gets the job done using the least time and effort while avoiding the unintended consequences. If the rationale and methods are efficient, they are also effective. However, the reverse is not necessarily true. In practical applicants, efficient handling means there is minimal stress to the horse, and coincidentally to the handler. Discussions of techniques that use "pressure and release" are more like a "go or no-go" than slow and steady. Descriptions of "psychological pressure" are hypothetical. Only if the human is actually touching the animal is "pressure" an accurate term. Efficient techniques applied when physically touching livestock and selectively touching horses in specific areas are as important as is the method and manner of approaching livestock.[11] The release component of the "pressure and release" philosophy should be replaced by the more efficient "reduction" of the stimulus instead. "Release" of a stimulus implies "to set free." It is a total release of a given stimulus in a relatively short time. Releasing the stimulus eliminates the ability to manage a horse's behavior continuously. Release can be interpreted by the animal as a lesson to move a few steps and then stop. The handler may not have meant to teach the horse this unnecessary start–stop pattern, but he did. Reduction of the stimulus allows the handler to continuously manage the stimulus presence and intensity. The horse's response to a stimulus is directed by uninterrupted human action, and that results in the horse's reaction being directed rather than corrected with no interruption. That is efficient.

The basis for efficient handling is that micromanagement of the human presence and action is used to continuously direct animal reaction, and a humane, intended outcome is the result. The intensity of the stimulus is managed or specifically directed and uninterrupted. Movement and learning are controllable and manageable. Efficient handling gives more value for the effort than effective handling does because horses learn from every interaction with humans. The efficient handler will have to devise methods that prevent an escaping horse from doing so or retrain the escape-prone horse. Retraining is almost always more difficult to accomplish than is preventing the problem behavior in the first place.

What Horse is Being Handled?

While some people advocate that bad horses do not exist naturally and are just products of bad experiences, that thesis is dangerous. There are truly dangerous animals. That fact needs to be kept in the back of one's mind. In addition, the size, age, and reactivity of the individual horse are important. These matter because mass and velocity have always been factors in safe animal handling, as are past behaviors, current reactivity, and strength. Fortunately, the size of the handler is less of an issue because strategy, thought, timing, and cleverness can be used for safe handling for the majority of horses.

Approximately 80% of backyard horses can be handled safely and efficiently with a variety of techniques if simple precautions are taken (Figure 5.2). The remaining 20% of animals are young, naïve animals, innately reactive horses, and horses that have developed maladaptive behaviors. The goal of any handling application should be *primum non nocere*—above all do no harm. Observing people handling horses suggests that about 95% of them correct a horse's behavior, and only 5% direct the behavior instead. Maladaptive behaviors can arise from any handling scenario and that includes veterinary procedures. Once developed, maladaptive behaviors are difficult to correct and control. One does not learn anything the second time he is kicked by a mule. It just hurts. In other words, retraining is very difficult, so learning, observing, and prevention are important. In veterinary practice or during routine handling procedures, startle and pain can be difficult to avoid, but the practical causes and the reactive effects can be managed.

There are a number of questions that need to be considered in order to determine how best to handle any specific horse. The answers may not be known for some of them and other answers are obvious. "What horse" means what is the age, breed, sex, and fertility state? Has this animal ever been haltered, led, and tied? Has the animal ever been vaccinated or experienced a needle? Does it react violently when attempts to handle or restrain it failed? Who is the primary handler of the horse and were they involved in any failed attempt at restraint? What time of day is the horse being handled? A warm, sunny afternoon during

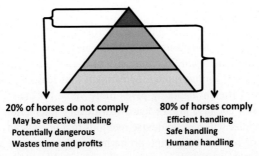

20% of horses do not comply   80% of horses comply
May be effective handling   Efficient handling
Potentially dangerous   Safe handling
Wastes time and profits   Humane handling

FIGURE 5.2   About 80% of horses encountered can be handled. It is the 20% that are most likely to impact safety, efficiency, and perhaps their humane handling. © *Don Höglund (2015).*

limited environmental activity, especially feeding, is a prime time to handle reactive horses. What person handles the horse from pasture to pen or from stall to stall, and how are these moves done? If the horse is currently reactive, is that handler present?

What Needs to be Done?

The type of procedure that is planned, from ear clipping, to hoof trimming, to a veterinary procedure, involves preliminary preparations. Does the procedure require that the horse is going to be touched, especially touched on the head, ears, or on the legs? These are particularly sensitive areas that are guarded by horses.

Animals learn to associate the motion immediately prior to startle or pain with the actual happening. After that point, any motion resembling the previous one can result in a negative behavior. The stress response will continue internally for at least 20 min. If the event is significant to the horse and reinforced by rough handling, startle, or pain, it is probably learned for a lifetime.

When the horse's response results in the handler stopping or withdrawing, the horse will quickly associate its reactivity with the withdrawal. The horse's unwanted reactivity was rewarded. To prevent this, it is important that the handler not stop what they were doing during the horse's reactive phase. If the horse becomes reactive to a handler's touch, the touching should continue safely until the horse stops reacting. In this manner, standing still will then be associated with handler withdrawal. The lesson is repeated until the horse has habituated to the touch. The point is that a reactive horse learns to be reactive if the handler withdraws during the reaction.

If the horse is reactive or becomes reactive, avoid reinforcing or repeating the same motion that was involved in creating the animal's reaction. The most vivid example is in the motion just preceding a needle penetrating the skin such as tapping the neck first. Motion becomes the precipitator of the painful event. The horse soon learns to associate a syringe and motion followed by pain. Instead, the solution to the "needle-shy" horse is to change the motion just prior to the needle penetrating the skin. The handler should rest his hand on the site to be injected and wait for the horse to stop its reaction. Then smoothly penetrate the skin without a jolt. The nerves are in the skin, so it is best to move quickly through the skin. In practical terms, do not repeat a motion similar to that which caused the horse to escape or avoid the injection. Escape is innate and avoidance is learned.

Modern horse handling has often promoted the need to continue applying a stimulus that causes reaction until the animal has habituated to it. More often than not, this does not work because the lesion is stopped too soon. Habituation requires that the lesson continue for however long it takes for the horse to give up. If not, the reaction is reinforced instead.

Efficient handlers incorporate predictable motion when handling animals. Slowing or removing quickness takes practice. It is also best for handlers to keep the movement of arms, instruments, and equipment below the horse's eye level.

Though many veterinarians have practiced successfully in an era where sedation was limited or rarely used, veterinary procedures on reactive horses should include sedation. This is especially true if handling involves the horse's head. Sedation makes these procedures safer and easier, and maladaptive behaviors are less likely to be learned.

WORKING WITH REACTIVE HORSES

Some horses are stoic and others are reactive to the point of being dangerous. A horse that is reactive at a distance will also be reactive if cornered. The reactivity trait tends to remain even as the animal gets older.[7] That presents an unsafe environment for everyone involved. It is a responsibility of the handler to interpret what environmental stimuli precipitate what horse behavior. Some efficient methods can help reduce this generalized reactivity over time. Likewise, there is a level of a stimulus where even tame animals become reactive. Handlers are well served to take a few moments and actually identify and count the stimuli that could be involved in promoting reactivity in the horse currently being handled. If stimuli (i.e., tractor noises and motion, barking or active dogs, loud humans, and feeding of nearby animals) can be removed, do so. The more astute the handler is at deciphering the cause and effect of stimuli, the more efficiently the handler can manage horse reactivity. Calm handlers usually have calm horses, and calm horses are usually safer to handle.

It is important to assess the initial behaviors when first interacting with a horse. Is the animal willing to stand or allow a human to approach or does it try to move away? Is it reactive and constantly scouting its surroundings, or calm (Figure 5.3)? If the horse is reactive, remove environmental stimuli and avoid adding energy to a tense situation. That is where aversive handling,

FIGURE 5.3   This horse is directing its ears back, indicating that the animal's attention is to the rear. If something startles the horse at this point, the animal may rush forward, creating safety issues. © *Don Höglund (2015).*

avoidance behaviors, and injury usually start. The approaching person should step back and away from the head. Give the animal time to calm down. Then a slow approach is used to help establish familiarity. Feed can be disruptive, so the horse should be moved away from any food-associated area.

Trapping

In order to handle the horse efficiently, it needs to be restrained. Chasing the horse until it fatigues teaches the horse to run from human presence. Trapping the animal in a holding area can be efficiently done by using calculated approaches of the horse's flight distance and controlling the environmental stimuli as much as possible. This can be done safely and with low reactivity. Throughout the procedure, the handler should watch the horse's shoulders and feet so that they do not stare at its eyes. Animals quickly learn from prior experience that when a human is staring at them, startle and pain are associated.

As previously discussed, the flight distance is the point where an intruder stimulates the animal to move away. If the approach is slow, the movement away is often slow. Handlers need to remember that horses cannot see directly behind them. If the person cannot see the horse's eye, the horse cannot see the person. That means an approach from behind is more apt to startle instead of caution the animal. Movements that are less than 18 inches (0.5 m) from the horse's nose are also likely to startle the animal (Figure 5.4). The ability to focus on these objects, especially moving ones, is difficult.

To direct the movement of reactive horses, the handler begins by walking parallel to the horse but in the opposite direction (Figure 5.5). Soon, the handler can angle to face the horse and then angle to move parallel with the horse's

FIGURE 5.4   The handler is standing just outside the horse's strike distance. If she moves her hand closer, it will be too close for the horse to be able to focus well, and the horse could step back or potentially strike out at her. Note that the ears indicate the horse's attention is directed at the handler. The handler is maintaining contact with the lead rope, neither pulling nor releasing, which allows her to manage the behavior through immediate prompts. © *Don Höglund (2015).*

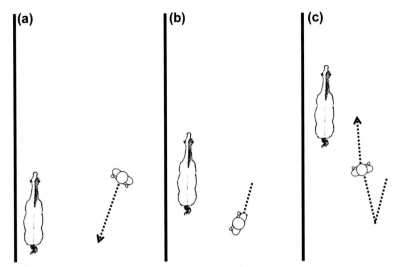

FIGURE 5.5   To start a horse moving forward, (a) the handler moves from the head toward the rear, with the distance between the horse and the handler determining the speed of escape. (b) Once the horse is moving forward, the handler turns to face the animal when the horse's hip is opposite the handler. (c) Then the handler begins walking parallel and in the same direction as the horse, staying no farther forward than its shoulder and no farther back than its hip. © *Don Höglund (2015).*

movement, staying between its shoulder and hip (Figure 5.6). This movement will usually help guide a horse from pasture to pen or from pen to stall.

Over the years, horse trainers have adopted many, varied methods for corralling and restraining horses. Many are effective. Some horse handlers choose to rope the horse; others chase the horse endlessly until the human or animal fatigues. Still others circle the horse continuously in an attempt to convince the horse that stopping is the solution to circling. In this case, the horse never gets any further away from the human and will eventually try some other form of escape. Some horses attempt to jump fences. Some stop and face away from the handler, and some face the handler. Still others use aggression as their defense. Running in circles in a round pen can ultimately cause the horse to switch directions, stop, turn, and perhaps face the stimulus to figure out what to do next. During this phase, the horse tries doing nothing or may use aggression. These methods are not efficient. They are time-consuming, can be associated with human or horse injury, and leave the horse prone to escape behavior.

Trapping with horse panels and adequate space is both efficient and effective. In this case, trapping does not involve chasing a horse into a small trap (as often shows up in the movies). It involves slow but definitive movements that encourage the animal to gradually move in the direction the handler intends and directs the horse into a small catch area. Because the movement is slow and calm, the negatives previously described are avoided. Trapping in this case begins the process of habituating the animal to human presence and touch.

FIGURE 5.6   The handler circles with the horse, positioned between its hip and shoulder. Other environmental stimuli are limited by the type of fence used and the handler's arms, whip, and direction of gaze. The handler's shoulder is pointed at the horse and he is circling in the direction of the horse's movement. Notice the right ear of the horse is directed at the handler, indicating that the horse's right eye is watching him. This is an excellent example of minimal handler stimulus and low-energy reactivity by the horse. © *Don Höglund (2015).*

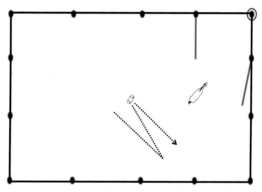

FIGURE 5.7   To herd a horse toward a trap, the handler uses a zigzag pattern behind the animal, being sure to go far enough to either side that she becomes visible to the horse's monocular field of vision. © *Don Höglund (2015).*

The trap needs to be a solid structure, such as that created with heavy duty, 10–12-foot portable horse panels. There are several configurations that can be used depending on when the trapping will be done, but it is important that at least one corner of the trap be chained to a solid structure like a stable pole or barn corner.

Either entice the horse into this trap area with feed or by herding it. When herding a reactive horse into the enclosure, the handler should use a zigzag pattern of straight lines, moving through, but well past, the blind area (Figure 5.7). Because arching movements hinder good visual perception, a horse tends to turn to look at the person moving, and this causes the horse to arc away from the desired straight line movement—thus the importance of the handler walking in a straight line. This zigzag technique can be used for trapping one horse or several.

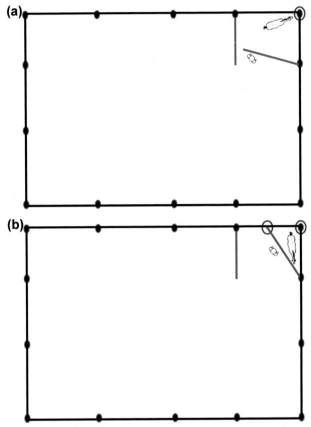

FIGURE 5.8   As the horse enters the trap, (a) one of the panels is swung to close off the side of the entrance, and the swinging continues past that point until it contacts the fence trapping the horse in a triangular area. (b) The swing panel is then chained to the fence. (Blue lines indicate panels and red circles indicate when chains are used to secure panels.) © *Don Höglund (2015).*

   When the single horse has been moved into the area where the horse panels are set up, the handler now has the opportunity to calmly move the animal into close confinement. This can be done in several different ways. Using the two-panel system in which they first form a small square (Figure 5.7), the handler will first close off the open side that the horse entered. He then continues the sweeping motion of the panel toward the horse until the free end of the panel touches the back fence, forming a small triangle that is just big enough to hold the horse (Figure 5.8). The free end of the panel is then chained to the fence to secure its position.

   If the horse must be trapped along a fence, instead of in a corner, two panels are connected and the free end of one of them is chained to the fence. That panel is extended out from the fence and perpendicular to it (Figure 5.9). The second attached panel is positioned so that it is perpendicular to the first panel and

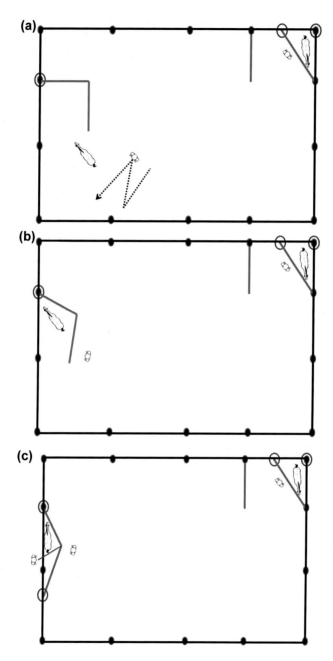

FIGURE 5.9   When a section of fence is used instead of a corner, (a) Two panels are linked together and set up so that they form an "L". (b) The horse is herded into the fence corner of the "L" and the panels rotated so that the free end of the set is moved toward the fence. (c) When the free end touches the fence, it is secured with a chain and a rope is attached near the connection of the two panels so that the panels can be pulled tighter toward the fence. (Blue lines indicate panels, red circles indicate when chains are used to secure panels, and a red line indicates a rope.) © *Don Höglund (2015)*.

parallel to the fence, forming a three-sided box into which the horse is herded. Once the animal is within the panel area, the free end of the panels is dragged toward the fence to trap the horse within a triangle. A rope is placed where the two panels join and is pulled by a person outside the pen to help move the panels toward the fence and tighten the space available for the horse. The free end of the panels is then chained to the fence to secure the trap.

A third option for trapping a horse can be used if the animal is inside a round pen (Figure 5.10). As with the straight-sided fence technique, two panels are joined and one end is secured to the side of the round pen. When the horse has entered that three-sided box, the free end of the panels is pulled toward the side of the round pen. A rope is attached to the junction of the two panels and is pulled by someone outside the round pen to help draw the panels closer to the horse. The free end of the panel completely closes the trap when it is chained to the side of the round pen.

Round pens can be used in another way to trap a reactive horse (Figure 5.11). Panels are set up in a pasture to create an alleyway that narrows as the horse approaches the round pen, aiding in moving the animal to the desired location. Once the horse is in the round pen, a panel is secured across the alleyway to block the return to the pasture. The horse is then herded from the round pen into the smaller trap and secured within the trap when the round pen gate is closed.

If a reactive horse is in a stall, a single horse panel can be used to trap it (Figure 5.12). Here, the horse will be confined against a stall wall that does not have any projections (like a water bucket or feed trough), and the stall door is to remain open to allow a safe exit for the handler in case of an emergency. The panel is slid in through the stall door to trap the horse in a portion of the stall, and it is important that the handler direct the panel from the end, not from the middle. One end of the panel is secured with a chain and the free end is rotated toward the wall closest to the horse, forming a small triangle to trap the animal.

Regardless of the type of trap used, once the animal has calmly entered, it is allowed to adjust to the confinement for several minutes before anything else is done. Now the handler begins approaching the animal using approximately 45% angles, but never passing the animal's hip or point of the shoulder (Figure 5.12). Horses are less reactive if approached at an angle rather than straight on. This forward movement is slow and calm even if the horse reacts. If the horse stops reacting, the handler stops and takes a small step back, ideally for about 30s. When the reaction starts again, the handler changes directions and continues encroachment. It is important not to stop the forward movement if the horse is reacting because that will reinforce the reacting behavior. The forward movement should end when the handler reaches the edge of the trap, and if at that time the horse stops reacting, the handler should reinforce the quiet behavior by taking a small step back. If it is reacting, continue a slight left-to-right rocking motion until the reaction stops and the handler can step back.

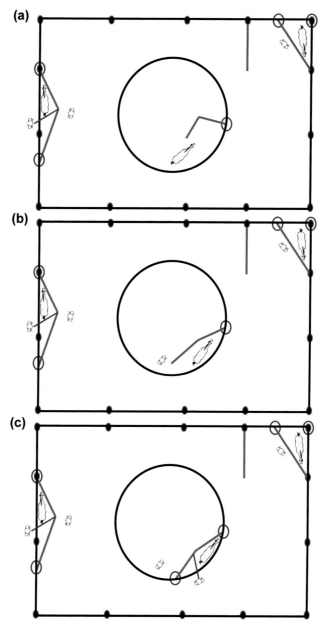

FIGURE 5.10   When a round pen is used to trap a horse, (a) two panels are linked together and set up so that they form an "L". (b) The horse is herded into the fence corner of the "L" and the panels rotated so that the free end of the set is moved toward the fence. (c) When the free end touches the fence, it is secured with a chain and a rope is attached near the connection of the two panels so that the panels can be pulled tighter toward the fence. (Blue lines indicate panels, red circles indicate when chains are used to secure panels, and a red line indicates a rope.) © *Don Höglund (2015).*

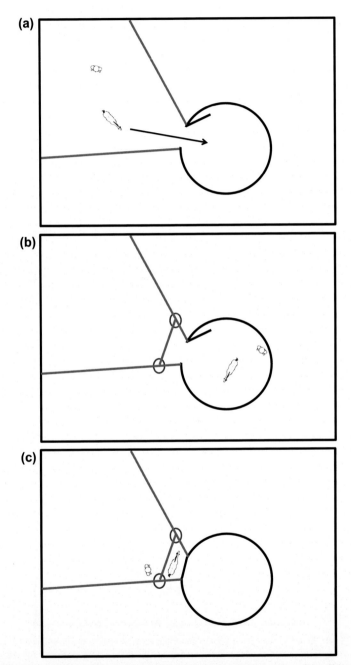

FIGURE 5.11   Another way to use the round pen in the trapping process works well on reactive horses. (a) Panels are used to set up an alleyway that narrows as it reaches the gate to the round pen to help steer the horse into the pen. (b) Once the horse is in the pen, another panel is secured to the alleyway panels, creating a trap. (c) The horse is herded into the trap and the round pen gate is closed to complete the trapping. © *Don Höglund (2015).*

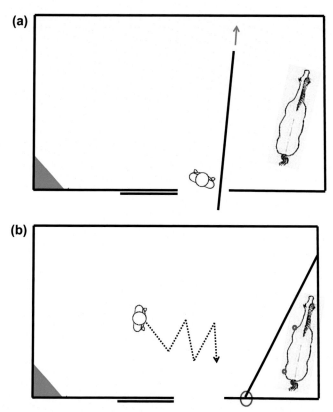

FIGURE 5.12 Trapping can be done in a stall by (a) sliding in a panel to block the horse between it and the side of the stall having no obstructions. (b) Once in the stall, one end is secured by a chain or rope to the stall and the free end is pushed toward the stall wall to form a triangle containing the horse. When the handler approaches the horse, he uses a zigzag pattern that is limited by the point of the horse's shoulder and its hip. © *Don Höglund (2015).*

Handling the Trapped Horse

Now the situation is right for an approach to begin. It is most efficient to approach the shoulder of the horse standing at a 45° to the horse. That is less of a stimulus than is a straight-on, perpendicular approach. If the horse rears, change the angle of approach to less than 45°, so the handler and horse are closer to facing the same way. Breathe normally. If the horse sighs—a sign of relaxation—the person should sigh with it.

Handling begins, not by touching, but by keeping the person's hands to himself until the horse smells the person. Humans avoid unfamiliar creatures or people who attempt to touch without proper introduction. This is true for horses too. A person begins the touch introduction by remaining at an angle to the horse near the shoulder, facing forward, and then touching anywhere on the side of its body that is below the horse's eye level (Figure 5.13). Do not pat or

FIGURE 5.13 When finally touching the horse, use the back of the hand to minimize points of contact. Touch the side of the body at a level lower than the horse's eye. Notice that the horse's left ear is pointed at the handler, indicating that the person is a focus of attention. © *Don Höglund (2015).*

pet. Because these are not things wild horses do or come in contact with, they are not understood. Association with good outcomes has to be learned. Touch with the back of the hand or a closed hand for one point of contact, instead of several points caused by each finger. If the horse reacts, do not take the hand away because it rewards the reactivity. Wait until horse lowers reactivity and then remove the hand.

Vocal prompts can be associated with human presence and action. Using a single syllable word vocal prompt, such as the word "touch," begins the process of associating the actual sound with the physical touching of the horse. When the handler is standing at the shoulder of the horse, a "touch" can be used to signal to the horse that he is about to be touched with the back of the hand within a second. Touch and hold for at least 30 s. Repeat the process several times, removing the hand only if the horse is nonreactive. If the process becomes unsafe, stop the process and start over. This process creates a predictable, seamless, smooth pattern where the word "touch" is followed by actual touch. Repeating the word "touch," or adding a clicker sound or a smooching sound complicates the pattern that the handler is attempting to establish. Give the prompt one time and follow it immediately with the physical touch.

Touching the head of the reactive horse is similar to touching other locations on the body, but touching the head and legs is more likely to cause a reactive response. Horses have very limited visual contact on their head, and their legs are associated with escape behaviors. Move slowly, be consistent, and purposeful. To touch the head, begin by touching the neck with the back of the hand for a few minutes and then move seamlessly to the cheek. Continue sliding the hand forward to the nose, and then hold the back of the hand against the nares, one side at a time. If the horse reacts, hold the position until the reaction stops. Remove the contact pressure when the horse stops moving. Once the horse has been conditioned to being touched on the nares, move slowly toward a spot under the eye. From there, the hand moves up to the ear, without grabbing it. It is safest to work with this technique across a stall door if the horse is considered to be dangerous.

After the handler has established contact with and can safely touch the horse without the reactivity, haltering is the next order of business. The halter to be used should be made of flat nylon straps or leather and have metal rings and a buckle. Rope halters and those with quick release clips should not be used. It is important that no part is easily breakable. With an assistant positioned on the opposite side of the horse from the handler and under the handler's direction, the halter strap that eventually goes over the top of the neck is first slipped under the neck of the horse and handed to the assistant. The handler holds the buckle in a manner so that his fingers are safe. Both individuals simultaneously, slowly, and safely move the halter forward under the horse's head. Next, the noseband portion of the halter is slowly slipped over the horse's nose and held there to await the reaction of the horse. After the horse settles, the handler should touch the horse's neck after using the prompt word and move seamlessly to the top of the neck to retrieve the strap from the assistant. The strap is then buckled snugly. No attempt should be made to begin by slipping the noseband over the nose first and then slipping the head strap past the eyes and over the ears. This is too much stimulus and usually results in the horse becoming dangerous.

TEACHING THE HORSE TO LEAD

A number of techniques have been used to teach a horse to lead, and techniques vary depending on the reactivity and size of the horse. Pulling the head with a lead rope attached to a halter is generally not very successful. Horses tend to balk, and then if they do go forward, they tend to jump. This makes the technique potentially dangerous to the handler. It also emphasizes the importance of standing to the side of the animal when leading it and being in a position to watch the horse's movements.

Once the horse has a halter on, it will need to become accustomed to receiving head pressure. The horse may drag a long rope that is thick enough so that it cannot self-tied around the horse's leg, and the animal needs to be kept in a small pen without objects the rope could snag on or entangle with. The tugs felt when the horse steps on the rope teach it not to panic and to give its head rather than fight. After about a week, the long rope can be traded for a small version that is about 12 inches (30 cm). This is used more as a handle for someone walking up to catch the horse so that their fingers do not get trapped by the halter. Catching the short rope first gives the handler more time to snap a lead rope onto the halter.

Foals in particular respond well to having a rump rope used so that any pulling is more like a push. A loop circles the hindquarters from just in front of the hip to slightly above the hocks. Hard pulling can cause the loop to rise under the tail, and this should be avoided. Rump ropes also cause the horse to look back

FIGURE 5.14 A technique to teach a horse to lead after it is accustomed to a halter is to place (a) a rope loop around the chest, in the general location where a saddle cinch would go, with or without an underlying band. (b) The long end of the rope is run between the front legs and up to the halter. (c) It is then run between the halter and the jaw. © *Don Höglund (2015).*

while moving forward. That is also not desirable. Initially walking the foal with the mare helps it learn that forward motion is intended. Because the mare remains calm, so does the foal. Gradually, short trips in a different direction emphasize the lesson of being led. Using a zigzag pattern to maintain momentum can be helpful.

Another technique employs a rope loop around the barrel of the horse with the long end of the rope coming from the area of the sternum between the front legs. From there, the rope goes between the halter and the underside of the jaw, exiting the ring where the lead rope would be attached (Figure 5.14). The advantage of this method is that the stimulus is moved from the neck to the withers, controlling the whole front end. This method also teaches young horses not to startle to the cinch when later introduced to a saddle.

ROUND PENS AS A TRAINING AID

No matter the corralling, restraint device, or the horse's behavior, understanding how horses learn is essential for humane handling and conditioning. The primary function of any corralling system is for containment and protection of the animals. Some livestock containment or corralling is also used for conditioning or training animals to perform for handler needs. Corrals, ropes, lunge lines, wire or post fence lines, walls, and wide open spaces have all been used in the training and retraining of horses. How they are used affects the outcome.

The round corral is not new. More than 100 years ago, the coyote fence encircled and protected calves and lambs, and it doubled as a corral for "breaking" young horses or sorting groups of animals. The original design of the coyote fence was solid walling with upright cedar tree branches woven together with hemp or wire when available (Figure 5.15). The solid nature of the structure

FIGURE 5.15 A coyote fence consists of upright, thin posts tightly woven together to create a solid sided corral. The original purpose was to keep coyotes away from barns, because coyotes need to grip with their hind feet when climbing and they cannot do so on vertically directed poles. While there are other fencing options currently available to coyote proof an area, the coyote fence remains aesthetically popular and can serve as a practical horse corral. © *Don Höglund (2015)*.

FIGURE 5.16 Round pens can have open views or views can be limited. When horses can see outside of the pen, they are likely to react to stimuli from outside. These unnecessary stimuli interfere with early training and may even predispose the horse to attempt to turn back or switch leads. © *Don Höglund (2015)*.

kept vermin out. It also limited the animal's vision to the interior of the pen to minimize outside stimuli from interrupting the lessons (Figure 5.16).

Locomotion is affected by the diameter of the interior of the round pen. Thirty feet across is the minimum recommended size for 6-month-old horses. A 40-foot-diameter round pen is probably a more efficient diameter for groundwork conditioning, while 60 feet across is better suited for mounted conditioning. Perimeter walls can be vertical or slanted slightly away from the center although both have disadvantages. Vertical walls can permit the horse's shoulder, hips, or the rider's knees to collide with the fence. Slanted walls have been associated with horses that attempt to climb the sides. With use, the dirt or other surface material of the round corral tends to build up along the foot of the wall, which does help move the horse and rider away from the wall during circling. Gates of the entrance and exit are best designed where they can close and latch from the interior of the corral, making them less likely to burst open if the horse collides with them.

Problems can be associated with any corral design, not so much from the design itself, but more as a result of the handling. Corral design as a training tool should depend on how the corral is used. Before using any corral for containment, consider what horse is being handled, what outcome is intended, and what behavior the horse is currently exhibiting. Round pens may not always be appropriate. For instance, young, inexperienced horses that have never been in a round corral often collide with the wall. Reactive young horses can be found to switch directions without warning and can become dangerous to the handler during attempts at escape behavior. If the goal of beginning handling is to teach the horse to stand still, then chasing or fatiguing the animal is likely not going to result in the intended outcome.

There are four disadvantages to using a round pen. The first is that dizziness of both horse and handler can be a by-product of the circling. The second is the unavoidable creation of learned helplessness in the horse because it can never actually escape. The horse switches directions continuously in apparent attempts to escape, though it is never successful. The third disadvantage of the use of the round pen is that the uninformed handler assumes that the horse is facing the handler in an attempt to communicate with the human. Instead, the horse faces the handler to assess depth and motion due to their visual limitations. Lastly is the ease with which aversive stimuli can be applied. From the horse's perspective, it may appear that the handler is continuously chasing it.

The primary advantages of the round corral are that the design allows for sustaining animal movement and there are no inside corners where animals can face away from the handler. The circularity of the round corral is effective when used to rate movement, prompting gait changes, or moving in a specific direction while maintaining a specific gait or lead leg. The handler should begin the training session by determining what is to be accomplished and what the horse can learn from the presence and action of the handler. Unidirectional, sustained movement of the horse is one outcome that the continuous circling can help achieve. A corner of a rectangular pen is a place a horse could retreat to, and he will likely repeat it. That is inefficient for conditioning the animal to move, rate speed, stop, or turn and move on prompt by the handler. In addition, a horse in a corner can kick out, which makes it unsafe for the handler.

Escape or aggressive behaviors are natural equine behaviors. They may be unwanted, but they are normal, especially in horses that haven't been gentled. As these horses attempt escaping, they whirl around the inside perimeter of any corral. From the horse's perspective, circling might cause the horse to become dizzy and unbalanced. That unpleasant experience may then be associated with human handlers, making escape the horse's primary reaction to human presence. Forcefully chasing the horse in a circle also teaches the horse to run from human presence. Eventually, it would learn that nothing relieves the handler stimulus, so it attempts to climb the wall, switches directions repeatedly, or stops and faces the handler. This is counterproductive.

Negative reinforcement is the technique most often employed in successful equine training regimens. The handler applies a stimulus, the horse reacts, and the handler responds by reducing the stimulus. Handlers who employ consistent stimuli, good or bad, will have horses that quickly learn to associate handler presence and action with their behavior. In the wrong hands, the round pen is an excellent place to generate conflict behaviors. For the experienced handler, the round corral is superior for producing consistent, desired behaviors.

Methods for maintaining, stopping, and starting locomotion in the round corral vary with each handler. What is critical is the need for minimal, consistent, and controlled stimuli? At the beginning of each training episode, counting the number of stimuli that are affecting horse reactivity is a good starting point. Other horses, dogs, equipment, human noises, ambient noises, and moving objects are all part of the total stimuli on the animal. Minimize the total number. After all, the horse is going to respond to the most meaningful and intense stimulus. That needs to come from the handler.

Consistent human presence and action is the key to efficient round pen conditioning. That means entering the pen from the same location in the same manner at the same rate and angle. Consistent routines establish consistent patterns of interaction between the handler and the horse. Entering at different points and at different energy levels presents the horse with a novel or unfamiliar stimulus. So does wearing different clothes or headwear. Human voice commands are merely stimuli. Employ a stimulus when needed and remove it when it is not. Whistling, clucking, smooching, and talking to the horse are only efficient tools if the horse has associated them with some reaction. Otherwise, do not arbitrarily employ them until an association with them and some horse or handler activity or outcome has previously been learned.

When a reactive horse enters the round pen, handlers should leave it alone for an hour or so to investigate the inside perimeter of the corral without handler intervention. At the appropriate time, a handler should enter the pen with the least amount of stimulus possible, always keeping in mind the handler escape routes in case a fast exit is necessary. The handler should remain backed against the gate in full view of the animal until the horse has had an opportunity to investigate the perimeter of the corral again. If the horse is excessively reactive, it may be necessary for the handler to exit the pen and introduce human presence to the horse from the outside instead. If the handler is experienced, the appropriate time to move to the center of the pen is indicated by the reasonable reactivity of the animal. A galloping or fast trotting horse does not indicate lowered reactivity and is the opposite of the intended gentling of the animal. If the reactive horse stops, the handler should immediately withdraw a few steps and wait for the horse to investigate and learn from the impact of the lowering of the energy between the handler and the horse. This is the hallmark of negative

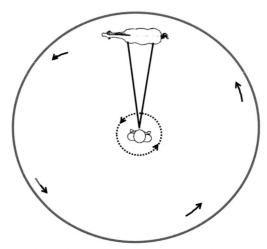

FIGURE 5.17 Once the horse will remain quiet in the round pen, the handler stands in the middle and turns in a circle with the horse, remaining in a line between the horse's shoulder and hip. © *Don Höglund (2015).*

reinforcement conditioning. Whips, flags, or other gadgets creating stimuli should be employed only when lack of reactivity or energy dictates.

Once the reactivity of the animal is at a safe level, the handler should move to the center of the round pen, facing roughly parallel with the horse as it moves. This begins the process of conditioning the horse to associate handler action with current horse behavior. The handler turns with the horse, albeit in smaller circles, staying positioned between the point of the horse's shoulder and the posterior aspect of the hip (Figure 5.17).

If the handler moves beyond the hip and circles from that position, the forward locomotion of the horse is difficult to manage. This causes the handler pattern to assume an oval-shaped circling, and that will ultimately cause changing encroachment of the horse intended to circle evenly (Figure 5.18). If the handler steps in front of the shoulder of the horse, this will cause the horse to move toward the wall and may cause the horse to reverse direction.

Circling with the horse and refraining from additional stimuli is most efficient when establishing patterns of association. A specific stimulus should be used for a specific horse behavior, and it needs to be applied consistently each time the horse behaves in that exact manner. For instance, if the horse walks in the circle, point an index finger of the trailing arm at the rear hoof of the horse. Turn slowly with the horse and try not to add any additional stimuli. If the horse speeds, step back a pace or two while pointing. This creates an association of the finger pointing down with horse walking and soon the horse will associate the position of the finger with his walking. Point that same finger at the horse's hip for trotting, and point the finger toward the horizon for galloping. For the unreactive horse, try using a stick with a small flag tied to the

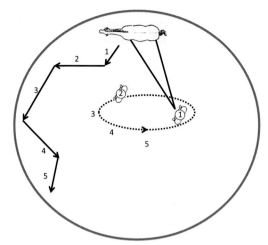

FIGURE 5.18   If the handler does not stay in a line between the horse's shoulder and hip, she travels in an oval instead of a circle and the horse responds in an erratic movement instead of a smooth circle because of the changing distances between them. © *Don Höglund (2015).*

end or a standard six foot lounging whip to replace the extended arm and finger. Additional stimuli such as handler motion, whipping, whistling, clapping, smooching, and barking dogs are confusing to horse. The association desired is where the horse learns that one prompt means one behavior. The handler must also apply only the intensity of stimulus that results in the appropriate behavior.

For excessively reactive horses, the handler should enter the pen when it is safe, and irrespective of horse responses, walk silently along the perimeter of the interior of the corral. In time this activity will slow the horse to a walk. The moment the horse slows, the handler should slow. If the horse stops, the handler stops and sighs. This process will teach the horse that when it slows, the handler stimulus will be reduced. The horse learns to walk in the human presence instead of run. This leads to stopping the horse, conditioning it to change direction without startle, and to stop on prompt.

Once the horse is walking either direction, as the handler circles the pen perimeter, the handler begins to make the circles smaller in diameter, until the handler is turning in the center of the pen. If the horse startles, the handler returns to the perimeter until the animal is walking again. Once the horse stops or faces, the handler turns and faces the horse. Facing the stopped horse and then moving toward it but at a slight angle instead of straight on will cause the horse to reverse direction. Working cow training of the horse prefers that the horse face the perimeter wall and turn (Figure 5.19(a)). This causes the horse to begin the use of the rear quarters for turning, as does the cow. Other disciplines of horse training may prefer that the horse face the handler at all times when turning to the different direction (Figure 5.19(b)).

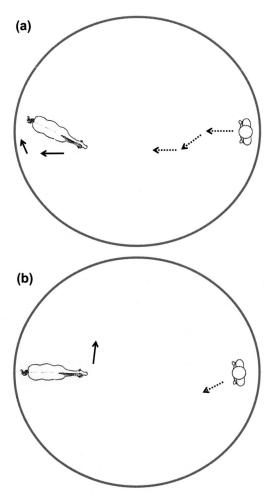

FIGURE 5.19   Handler movement toward the horse directs which way it will turn. (a) Disciplines that want the horse to turn facing the fence begin with the handler moving the head toward the wall, and then the handler moves toward the horse's head so that it turns to go the other way. (b) When the horse is to turn away from the wall, the handler moves toward the horse but away from the direction the horse is to go. © *Don Höglund (2015)*.

The time to stop a training session is predicated on the success of the session. If the highly reactive horse continues to race the interior of the round pen, the session is stopped and trapping is used to teach the horse to stand in the presence of a human. If the horse repeats a desired outcome three times out of five, it is probably time to move on to another exercise. If a prompt fails to elicit the desired outcome, try not to repeat the same exact motion or prompt. Change some small portion of the motion and try again. Once the horse will stand unreactive for handler encroachment and allow indiscriminate touching and manipulation of the head and limbs, it is time to move on to more advanced training.

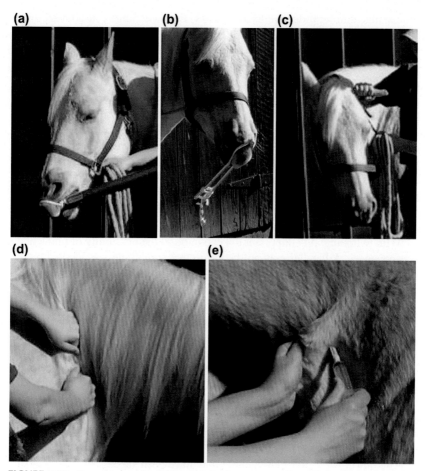

FIGURE 5.20   Several types of twitching techniques are commonly applied to horses. (a) Upper lip twitching can be done with a rope or chain loop twitch, and the twist should be rolled toward the ground (counter-clockwise if standing on the horse's left side) so that the lip is not pulled upward (b) The upper lip can be twitched with a clamp twitch (also called a humane twitch) (c) One ear is squeezed by a hand in the ear twitch (d) A skin twitch is done by squeezing skin with one or both hands. The head drops and turns toward the twitch side, making is easier to work on the opposite side of the horse, and ear twitching by hand (e) The one-hand twitching technique can be effective by rolling the skin to hide a syringe and change the horse's focus just prior to an injection. © *Don Höglund (2015).*

RESTRAINT TECHNIQUES

Various methods for restraining horses have a very long history of use. Common procedures include the upper lip, skin, gums, tongue, tail, or ear twitching (Figure 5.20). These applications of apparently painful stimuli during horse handling are commonly done for things like inserting a stomach tube, injections, standing castrations, bandage removal, clipping, farrier work, dental procedures, and reproductive examinations. Ear twitching is ill-advised under any circumstance because of the unintended consequence of creating horses that are averse to having their ears touched.

Why twitching works has been the subject of several theories and research projects. Theories have suggested that the twitch creates a "freezing" behavior in the horse that apparently overrides innate survival behaviors such as escape. Others theorize that the twitch is a distraction or that it "overshadows" the application of a second painful procedure.[6] Research has shown that lip twitching causes endogenous endorphin production. The subsequent sedative and analgesic effects take approximately 5 min of twitching to achieve effect, and then last 10–15 min after twitching.[6] Because twitching is typically an intense on-off phenomenon, studies are needed to determine how intense the actual twitching must be in order to achieve clinical effect. Twitching alone could theoretically interfere with lameness examinations due to the endorphin release.

With twitching, it is important to keep in mind the possibility of unintended consequences. A handler may not be aware that something was learned by the horse, but that does not mean it did not happen. Twitching in the wrong hands, is on a destructive course parallel with kicking a horse in the belly or grabbing an ear. Because animals learn from every encounter with their environment, any variety of twitching has true potential to teach a horse that handlers and devices are associated with head restraint and pain.

The use, duration, and intensity of the twitching event are controlled by the handler. One of the goals of twitching should be to create a minimally reactive horse during handling. Another goal should be to teach the horse to stand unreactive. That is no easy task, but it can be done. It requires patient practice of application, management of the stimulus, and well-timed reduction and release of the twitch. A final goal for twitching is to accomplish the original, potentially noxious task safely and efficiently.

The procedure requiring twitching may have been successful, but the result left the horse with iatrogenic and potentially dangerous behaviors. For that reason alone, twitching techniques need to be understood so that they can be applied and removed efficiently. Before any type of twitch is applied, it is advisable for handlers to evaluate the potential behavioral outcomes of twitching against the use of sedation and analgesia. If the handler and examiner do not want the horse to remember the event, short-term or reversible sedation and analgesia can be used, procedure permitting.[18]

Twitching the Upper Lip

Unusual smells that accompany pharmaceuticals, vitamins, and latex gloves, coupled with gatherings of groups of people or dogs, create new stimuli for a horse to process. If the situation is new to the horse, it will likely react to the changes irrespective of the presence of a twitch. The horse learns and likely remembers the critical environmental factors of motion immediately preceding the application of the rope or metal bars of the twitch and the pressure or pain receptor stimulus that directly follows the motion.

Slow handler movements toward or away from the horse's nose are usually more efficient than quick-grabbing hands that immediately clamp onto the upper lip. It is important to remember that inside of 18 inches (1/2 m) of the horse's eye, their ability to focus is limited. Fast movement near the head of the horse can cause startle or escape behavior irrespective of the presence or absence of a twitching device. Touching the horse's neck and seamless movement of handler hands up the neck and along the cheek toward the nose is a tried and true method for approaching the horse's upper lip in a predictable manner. Even then, horses experienced in twitches often react to the mere sight of one. Experienced handlers become good at being able to keep the twitch out of sight until it is actually applied.

People who are applying the twitch should stand on the same side of the horse as the person working on the animal. The head can remain pointed straight forward or turned toward the side. In this latter technique, a handler standing on the left side of the horse grasps the halter cheek bar with the right hand while also holding the lead line in the right hand or draped over the right arm. Then the handler steps back toward the point of the horse's shoulder, gently pulling its head toward the shoulder until it is 90° from straight forward (Figure 5.21). This places the handler away from a direct strike by the front legs (Figures 5.22 and 5.23). If the horse insists on circling around the handler, align the horse against a solid wall or sturdy, wire-free fence line.

Rope loop twitches are preferred to chain loop ones because there is better control of the stimulus pressure. At some point, a twisted chain will twist

FIGURE 5.21   The person who is going to twitch a horse can begin with the horse's head pointing straight forward, or as in this figure, turned to a 90°angle immediately prior to applying the lip twitch. © *Don Höglund (2015)*.

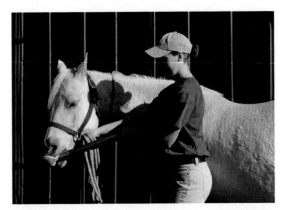

FIGURE 5.22   The person handling the lip twitch should stand to the side of the horse at its shoulder to avoid being injured if the horse should strike or rear. The handler is also controlling the lead rope and cheek bar of the halter. © *Don Höglund (2015)*.

FIGURE 5.23   Twitching a horse can be dangerous, particularly so if the person stands where they can be struck by the horse's front foot or hit by its swinging head. The handler should be standing by the animal's shoulder and have control of the halter as well. © *Don Höglund (2015)*.

no further and start to kink. This can cause secondary pain by pinching the entrapped skin.

The twitch is held in the left hand with the twitch's rope loop draped between the fingers and the thumb. The handle end of the twitch is held between the thumb and index finger of the right hand. This allows the twitch handle to slide along the thumb and index finger.

Slowly advance the hand with the rope loop along the right arm toward the halter. Then, touching the horse's cheek, advance below the nostril of the horse,

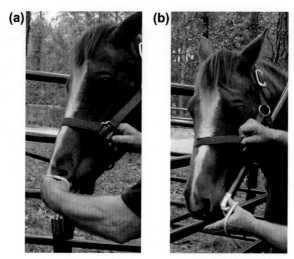

FIGURE 5.24 Once the handler has the loop of the twitch draped between the fingers and the thumb of the left hand (for a procedure on the left side of the horse), that (a) hand is gradually moved along the cheek until the upper lip can be grasped by the fingers. (b) The handle of the twitch can be left to dangle or secured between the halter and head so that it will not accidently swing and spook the horse. © *Don Höglund (2015).*

and then slowly grasp the upper lip with the fingers of the left hand (Figure 5.24). If the horse reacts, merely hold the hand on the lip and do not grasp. If the horse attempts to bite, grasp a full hand of lip and squeeze. Do not allow the horse to bite the handler or examiner.

Once the lip has been grasped, slide the rope over the fingers and well onto the lip. Experienced handlers twist the handle of the twitch so that the lip turns down instead of up, until it is tight around the lip. If standing on the left side of the horse, twist the handle of the twitch counter-clockwise. Slow increases in twitching tightness are usually more humane than immediate, full-pressure clamping. Stand back toward the point of the shoulder of the animal, and watch for trembling of the eyelids and lips as a sign that the twitch is too tight. If the head begins to droop, the endorphin effect is beginning.

Lip Chains

The lip chain is the most controversial of twitching methods (Figure 5.25). Careful use can teach a horse that moving means gum pain (if the only pain is associated with the horse moving), but in inexperienced hands it can be damaging to the gums and abusive. If the horse pulls back or breaks loose, stepping on the attached rope, it can significantly damage the gums. Future attempts to place a hand or a bit into the horse's mouth can be very difficult. This technique is commonly used in the race horse industry and on brood farms.

(a)   **(b)**

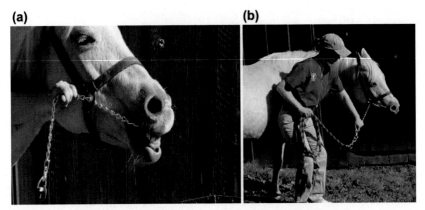

FIGURE 5.25 The use of the lip chain (a) on the gums is controversial because it is very easy to accidently cause a great deal of pain for the horse. (b) With the handler pulling down smoothly, without a jerking motion, the head can be lowered. Jerking on the chain will cause the horse to pull abruptly upward. © *Don Höglund (2015).*

Ear Twitching

The ear twitch is another type of twitching technique that can be associated with significant problem outcomes. The hand squeeze alone is not sufficient pressure to cause endorphin release. For proper effect, the handler should place the elbow against the horse's neck, and the ear should be pulled out and down while being squeezed. This process should be continued for 5 min. The illusion that handlers are causing sedative endorphin effects in a few moments of ear twitching has not been validated by research, but there appears to be some distraction from the pain if done properly. It can be sufficient to allow the passage of a stomach tube or allow a quick eye exam. More often, though, ear twitching causes more problems than benefits. In the age of short-acting sedative and analgesics, the use of the ear twitch should be discouraged.

Skin Twitching

Skin twitching takes strength and stamina while standing in a relatively dangerous area near the front legs. The technique does have some effect, but it is not likely to be related to the endorphin release. The procedure may be effective more as a distraction from other things than as a true restraint technique. Horses often become aggressive following excessive or the prolonged use of twitching, probably as nerve fatigue develops. Blinking; lip, muscle, and leg trembling; groaning; multiple sighs; head shaking; and abrupt escape or aggressive behaviors are indications of increasing nerve fatigue. At this point, the animal is no longer being influenced by the twitch. One way to prevent this is to periodically release the twitch application completely, meaning the handler will to have to reapply it again. It is more efficient to increase the stimulus and reduce it (tighten and loosen) as needed for specific painful portions of a procedure.

ENTERING AND EXITING A CONFINED SPACE

The literature is replete with foolproof philosophies and methods used to encourage a horse to enter a space and remain there. The space could be a trailer, float, chute, stall, scale, or narrow alley. Some techniques employ positive reinforcement (food), and others rely on well-timed negative reinforcement. However, too many rely on punishment (startle or pain). Trailer and chute loading applications may combine positive and negative reinforcement methods to load spaces efficiently, with some being more efficient than others.

There are as many ways to force a horse into a space as there are handlers with an imagination. An understanding of how horses learn and which behavioral principles are likely involved will help tremendously. In one study, 92% of nationally accredited horse riding coaches misunderstood the use of positive reinforcement, 90% misunderstood punishment, and 78% misunderstood negative reinforcement.[14] Misunderstanding learning theory, explaining altruism or meanness as the reason for equine responses, and assuming that horses are willing participants that comprehend human intention are likely causes for the problem. The absence of avoidance behaviors (rearing, bolting, kicking, striking, and biting) should be an objective of all horse loading procedures.

Escape behavior in horses is a naturally occurring survival behavior observed as a defense to novel or intense challenges. Avoidance behavior is learned. Evaluating which equine behavior the handler is encountering begins the experience of conditioning a horse to enter a confined space. Elimination of escape behavior requires training, beginning with familiarizing the horse to a space that may be new to him. The elimination of avoidance behavior requires retraining often without knowing the initial event.

Teaching a horse to load in a trailer or a chute has a significant potential to produce unwanted behaviors. For instance, circling the horse in front of the entrance teaches the animal to face away from the objective. If the horse bolts from the entrance or is circled to reface the entrance, escape behavior and learned avoidance combine to make entering the space a problem. If the horse is allowed to turn away from the objective, he will likely learn that he can turn away.

Devices like whips, lip chains, sticks or poles, and electric prods are the tools employed by handlers intent on correcting phantom "attitude" problems in horses. When a handler comes up from behind a horse and pops it with a whip, logic should say that the pain will quickly become associated with the trailer, the area around the back of the trailer, the handlers, handler motion, and the loading process in general. Such devices should not be present. That eliminates any temptation to use them.

Positive reinforcement avoids aversive stimuli by its very application. Positive reinforcement, such as food, can be valuable. If the reinforcing food works, use it, but place it in a pan or bucket to avoid teaching hand biting. Once the loading behavior is well established, reduce the amount of food with each repetition to eliminate the long-term need for it. When positive reinforcement for loading fails, use appropriately timed negative reinforcement to accomplish the task. It is important, though, to be prepared to immediately transition from positive to negative reinforcement.

Whether a horse is learning to enter a space that is unfamiliar to the animal or a space the horse has experienced before, the handler should always prepare for escape or avoidance behaviors. At the other end, there can be problems too. Teaching a horse to exit a stall, chute, or trailer slowly is often as complicated as teaching the animal to enter the space. The handler should prepare for the exit at the same time he prepares for teaching the horse to enter. If a handler is teaching a horse to enter a trailer with no front handler exit, both the handler and the horse need to somehow exit the compartment safely. Before entering any trailer, determine how and by what route both human and horse will exit. If training an inexperienced handler to trailer a horse, use an experienced horse to train the human. Likewise, use an experienced handler to train an inexperienced horse. If the handler cannot explain the process, including the exit, he is not ready.

Before beginning, examine the environment for potential stimuli of reactivity and remove as many as possible. This not only includes things like dogs and cats, but also means no intense lights or darkness inside the trailer.

If the horse is wearing a halter, it has been handled at least once. If the horse was led to the opening of any space new to him, he has prior experience being handled. To what extent the horse has been handled around a trailer or other confinement can be evaluated in mere moments. Animals showing reactive behavior when approaching confined spaces require more evaluation for the environmental or learned precipitators of the reactive behaviors. Where the horse directs the ears is an indication of what the animal may be evaluating through sight and sound. Use this information to determine if the horse is reacting to stimuli from in front of him or if the horse is reacting to stimuli emanating from elsewhere.

Though it is not always possible, as the old saying goes, "the restaurant is not the place to train youngsters about appropriate table manners." Train young humans and horses at home before expecting them to perform appropriately elsewhere. Patience with the young or naive animal is a virtue.

Teaching a horse to load or enter a space is most easily done if the space to be entered is wide enough to safely allow both the handler and horse going in at the same time. Side-by-side trailers will permit this if the center petition can be removed or the person can easily go into the adjoining stall.

Step one in loading is to keep the horse facing the destination. The lead line is intended to enable the handler to keep the horse facing the destination, not to pull the animal to it. If a trailer-broke horse is used too, it should be loaded into the front of the trailer. If it is occupying one stall in a side-by-side trailer, a long lead rope can be run up through the stall, out the escape or feed door, and either back to the handler or to an assistant. The purpose of the long line is to keep the head pointed toward the opening, so slack is removed as the horse steps forward.

Trips up to the space that familiarize the horse with the surroundings should be done slowly so the horse does not try to get away. Turning away from the objective can condition the animal that the pattern is to approach and then turn away from the destination. Circling in and around the objective has the same counterintuitive outcome. Pulling the animal toward the destination can teach it to expect intense pressure or pain upon entry. There is a fine line between enough pressure and too much. Limit the horse's options and the process has a better chance of success.

Every event leading up to trailering a horse helps predict how loading will go. Handlers should study what the animal tends to do when approaching gates or passing over objects, such as logs positioned on the ground. Just prior to attempting to load the reactive animal into a trailer, teach the horse to lead through gates with headers and footers as a method to familiarize a horse with stationary objects it must pass over and under simultaneously. If successful, move immediately to a prepared trailer. If the horse avoids moving through gates with headers, handlers can be aware that trailer loading is premature. Conditioning a horse to back off of trailers is likewise accomplished by teaching horses to back through gates with headers and footers and over a rail placed on the ground. If the horse will not walk through or back through a gate or stall door, it is likely going to exhibit avoidance behaviors while being loaded into and unloaded from trailers, floats, or chutes.

Training in segments, or "shaping" horse behaviors one segment at a time, has been successful at creating predictability among associations and contingency. If the horse refuses to cross the rail on the ground from any direction, forward or backward, it is time for ingenuity. Surround the horse with four posts on the ground, making a square around the horse, and place a small amount of his favorite food (positive reinforcement) some distance away. Sooner or later every animal will step over a post. If the horse refuses to step back, trap the horse in a "V" shaped trap so that backing is the only exit. Place a rail on the ground at the exit and use positive reinforcement (food) placed outside the trap as the lure. Soon, the horse will use trial and error to figure out how to back through the exit and over the rail. The horse should not be able to turn around once in the "V" trap, and the amount of food used is minimal so that several trials can be performed in the same session.

Familiarizing the animal with the space in the trailer, float, or chute happens by entering it. The handler should stop the horse in front of the opening for a brief moment or two to allow it time for a quick evaluation of the environment. Waiting at the entrance to the trailer for more than 30 s can allow avoidance behaviors to develop. Avoid stopping a horse that is loading without incident. A stopping motion can teach the horse to halt at the trailer entrance inadvertently.

FIGURE 5.26   The horse should be led directly to the trailer, and if there is no hesitation, the handler should move directly into the trailer without hesitation. The horse will continue its forward momentum and load without incident. © *Don Höglund (2015)*.

Creating and maintaining momentum are among the most valuable methods that handlers can learn (Figure 5.26).

If the horse attempts to escape, rush through, or avoid the opening of the gate or trailer entrance, every effort should be made to keep the animal from turning or backing away from the objective. The first time a horse avoids or attempts to escape the objective, the handler should try not to repeat the failure and subsequently risk reinforcing learned avoidance behaviors. If the horse is successful at escaping or turning away, that single event may be enough to train the horse to avoid openings. Therefore, the handler needs to prepare for all outcomes. Once loaded, hold the animal for more than 30 s on the trailer to teach standing and staying in position. A small food reward can be helpful to keep the horse calm. As the procedure is repeated, the horse is held in the trailer for gradually longer periods of time. Do not start training the horse if there is not adequate time to finish five successful events.

For the novice horse or horses known to be difficult to load, a rope circling the barrel where a cinch would be that then goes between the front legs and up through the halter helps ensure that the horse faces the opening at all entry times and that any tugging on the lead line is transferred to its body and not its head and neck (Figure 5.14). It is also important to position the trailer strategically. The trailer should be on level ground or slightly sloping so the trailer entrance is lower. It needs to be secured to a transport vehicle or to some other unmovable object. Lightweight horse panels should be aligned alongside the trailer or chute, and assistants should be ready to move the panels.

If the horse stops, maintain tension on the lead line but do not pull. If the horse backs, by virtue of his superior strength, go with him one time and force him to back until he stops. Stop forcing him to back when he stops voluntarily. Backing for horses is more difficult than forward locomotion and at some point they will stop. Keep the horse from turning by maintaining tension on the lead

FIGURE 5.27   If the first attempt to walk the horse directly into the trailer fails, assistants should move horse panels from along the sides of the trailer backward and begin to swing the outside ends toward each other until they meet and form a triangular-shaped trap around the horse. © *Don Höglund (2015).*

line so that the horse faces the handler and trailer while backing. This is an aspect of negative reinforcement that teaches the animal that when he stops backing, the handler stops motion too. Have assistants at the ready before making the second attempt at loading so that they can move the panels immediately. If the horse stands at the trailer entrance for more than a few seconds, that pause can become part of the pattern and the horse will attempt to stand instead of loading in subsequent loading attempts. Each assistant is positioned at the very ends of the panel to walk with the panels from both sides of the trailer in a direction parallel to but opposite from the loading direction (Figure 5.27). Horse panels are safer than livestock panels because they have no sharp edges.

As the horse moves forward, the panel assistants swing the ends of the panels together to entrap the horse directly behind the trailer entrance. The assistant by the trailer door can then slowly push the door toward the horse. Generally, the horse loads. For resistant horses, the end of the panel and the edge of the trailer door are angled together to reduce the space even more. It is important that if the horse were to kick out backward, all handlers are in a safe position. They are never to be directly behind the horse.

Exiting the trailer calmly is another important lesson for the horse. After the horse has remained calm in the trailer for a few minutes so that it learns not to bolt back out, the handler should turn the horse toward the door and walk slowly, one step at a time, toward the exit. The horse should stop at the exit and wait for the handler to initiate the exit. Handlers should step down and left, out of the horse's path (Figure 5.28). If the handler waits for a few moments, the horse will reach down with his front limbs, searching for the ground instead of jumping out (Figure 5.29).

Immediately upon exiting the trailer, the handler should guide the horse to turn around and face the trailer entrance. This conditions the horse to turn back and face the trainer and trailer instead of jumping off and darting away.

FIGURE 5.28   When teaching the horse to unload, the handler first has the horse stop at the edge of the trailer. He steps out and stands in a position to the side of the horse, so that if it should jump out, the handler will not be hurt. © *Don Höglund (2015).*

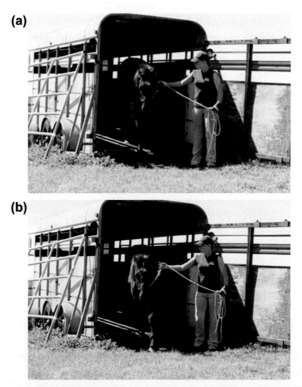

FIGURE 5.29   After the horse has stood at the edge of the trailer for a little while, it is gently encouraged to take a step down. By stopping at the edge, the horse has time to get its bearings and realize it can exit by stepping down, rather than having to jump out. © *Don Höglund (2015).*

(a)

(b)

FIGURE 5.30   As soon as the horse exits the trailer, the handler should turn it toward the entrance again. This teaches the horse to turn back and face the trainer and trailer instead of darting away. © *Don Höglund (2015).*

The turning should happen right next to the trailer opening (Figure 5.30). The loading and unloading process should be repeated with assistants ready with the panels. After several repetitions, panels will no longer be needed.

It is also important to switch handlers so that the horse generalizes loading with a trailer and not with a person. Different trailers and different locations increase the generalization of the intended behavior.

There are a number of other methods used to teach horses to load into trailers, and each has variations based on the reactivity of the horse and the type of trailer used. One technique is to put a trailer in a pen and place all food and water in the front end of the trailer. Another method involves trapping the horse in a box constructed with the panels behind the trailer and waiting for the horse to load of its own accord. Ropes placed around the horse's rump or people interlocking their hands have been successful, albeit less safe and more complicated for novice handlers and reactive horses. Young horses that lead well and know how to back can easily be taught to load into a horse trailer if an older, trailer-broke horse is loaded first. This takes advantage of the normal herding behavior of horses and does not represent observational learning. At the end of the process, give the horse a little time in the trailer to eat a snack before having it be the first to back out.

For trailers that require this on-forward/back-off approach, it is important that the horse knows how to back first. The backing-off system for novice adult horses is brimming with the potential for creating avoidance behaviors. Using aversive stimuli like lip chains to encourage a horse to back off of trailers because it does not understand what "back" means will likely teach the horse to associate being in a trailer with pain. The handler can back the horse one step at a time if he can stand near the horse's head. If the handler cannot be next to the horse, he can use a long lead rope so he can stand to the side of the horse's rear and yet give a gentle backward pull on the halter. For gentle horses, a useful technique to prevent the rapid exit is for an assistant to slowly back the horse, while the handler holds one hand gently against the horse's rump. Just as the horse is going to need to step down or step onto the ramp incline, the amount of pressure on the rump is increased. Abrupt changes are the most likely to trigger bolting behavior because they are unpredictable from the horse's perspective. After a few repetitions, the horse makes the connection that it will be signaled when the abrupt change is going to happen.

Though it is best to teach horses to load into trailers and chutes when young, older horses can and need to learn these methods too. The panels may need to be attached to the trailer and fashioned in a box structure for safety. In the beginning, it is better to have fewer people around.

When people stand behind the horse upon entry or use whips or brooms as negative reinforcement, the horse really is learning to look back or to expect that pain. The horse might load into the trailer, but the lessons do not involve humane handling.

OTHER USEFUL HANDLING TECHNIQUES

When working on the leg of a horse that is likely to strike or kick, an additional person can be particularly useful. With the horse standing on the leg, the helper can lift the opposite foot off the ground and hold it up. Horses typically try to free the held leg, so the helper needs to be experienced, but this does add a measure of safety. This may be sufficient restraint so that twitching is not needed. Another technique to pick up a horse's foot can be done safely by a single individual (Figure 5.31).

Horses can also be taught to lower their heads. While standing to the side of the animal and out of the line of fire for striking horses, the handler should clasp the cheek bar of the horse's halter. Then, slowly and gently, move the horse's head side to side in a one-foot-wide path. Simultaneously, apply force down on the halter. This downward force produces pressure on the poll of the horse's head. The moment the horse dips or lowers the head, immediately stop the downward pressure. This is using negative reinforcement to teach head lowering. Repeat the lesson several times until the horse readily responds.

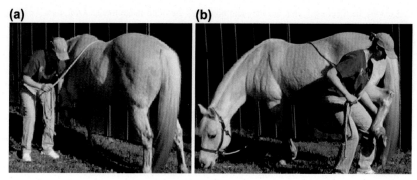

FIGURE 5.31 Single handlers can safely pick up the front and rear legs of a horse by using a wall to prevent movement away from the person. The lead is used to prevent forward movement and movement toward the handler by taking it from the chin ring of the halter, between the front legs, along the side opposite the handler, and over the back to the handler's hand. © *Don Höglund (2015).*

Horses are not born with an innate understanding of patting. At best it is a secondary reinforcer. It can eventually be understood if it is associated with a positive reinforcement like food. For this reason, smooth stroking with the direction of the hair is more effective.

People have a tendency to give too much credence to genetic predisposition and tend to ignore the individual's potential for learning.[11] On the other hand, overestimating a horse's mental ability is a welfare issue because it leads to waste, abuse, distress, and avoidance behaviors.[11] When a handler assumes the horse has an attitude problem rather than a bad experience, aversive correction methods are often employed. They make the mistake of thinking that pain will convince the horse of its wrongdoing. The horse then expresses avoidance behaviors whenever that handler is around, and a viscous, negative cycle begins.

REFERENCES

1. Anthony DW. Bridling horse power. In: Olsen SL, editor. *Horses through time.* Lanham, MD: Roberts Rinehart; 2003. p. 57–82.
2. Gilger BC. *Equine ophthalmology.* 2nd ed. New York: Elsevier; 2010. p. 429.
3. Goodwin D, McGreevy P, Waran N, McLean A. How equitation science can elucidate and refine horsemanship techniques. *Veterinary J* July 2009;**181**(1):5–11.
4. Gosling S. From mice to men: what can we learn about personality from animal research. *Psychol Bull* 2001;**127**(1):45–86.
5. Grandin T. *Humane livestock handling.* North Adams, MA: Storey Publishing; 2008. p. 227.
6. Lagerweij E, Neils PC, Wiegant VM, van Ree JM. The twitch in horses: a variant of acupuncture. *Science* September 14, 1984;**225**(4667):1172–4.
7. Lansade L, Bouissou MF, Boivin X. Temperament in preweanling horses: development of reactions to humans and novelty, and startle responses. *Dev Psychobiol* July 2007;**49**(5):501–13.
8. LeDoux JE. The slippery slope of fear. *Trends Cognitive Sci* April 2013;**17**(4):155–6.
9. LeDoux JE. Coming to terms with fear. *Proc Natl Acad Sci* February 25, 2014;**111**(8):2871–8.

10. LeDoux JE. Feelings: what are they and how does the brain make them? *J Am Acad Arts Sci* Winter, 2015;**144**(1):96–111.

11. McGreevy P. *Equine behavior: a guide for veterinarians and equine scientists*. New York: Saunders; 2004. p. 369.

12. McGreevy P. The advent of equitation science. *Veterinary J* November 2007;**174**(3):492–500.

13. McGreevy P. *Equine behavior: a guide for veterinarians and equine scientists*. 2nd ed. New York: Saunders; 2012. p. 378.

14. McLean AN. The positive aspects of correct negative reinforcement. *Anthrozoös* 2005;**18**(3): 245–54.

15. Miller RM, Lamb R. *The revolution in horsemanship: and what it means to mankind*. Guilford, CT: The Lyons Press; 2005. p. 368.

16. Moates T. A Horse's thought. *America's Horse* September 2013;**17**(1):56.

17. Morgan CL. An introduction to comparative psychology. London: Walter Scott Publishing; 2007. p. 382. Reprint of 1894 edition.

18. Shader RI, Dreyfuss D, Gerrein JR, Harmatz JS, Allison SJ, Greenblatt DJ. Sedative effects and impaired learning and recall after single oral doses of lorazepam. *Clin Pharmacol Ther* May 1986;**39**(5):526–9.

19. Tinbergen N. *The study of instinct*. New York: Oxford University Press; 1951. p. 228.

Chapter 6

Dairy Cattle Handling: Practical Applications of Science

In all affairs it's a healthy thing now and then to hang a question mark on the things you have long taken for granted.

Bertrand Russell

My grandfather kept a small herd of cows and a feisty bull which served the neighbor's cows, but we were never sure, as children, just what went on because we were sent to the house when a restless cow was brought into the barnyard. We asked Grandpa to teach us to milk, but he refused, saying the cows would give down their milk only for him, as they knew his touch. On the few occasions when Grandpa was sick and a neighbor or our Uncle Martin did the milking, we found Grandpa was partly right. The cows did give some milk, but not as much as with Grandpa. They were definitely partial to his touch. Once a cow who didn't like a stranger on the stool kicked over the milk pail and this did not please Grandma who depended on the cream for her butter-making. But Grandpa was never sick for long and soon the milk flow was back to normal. We girls still wanted to try milking a cow. Uncle Martin said he had a cow he'd let us milk. So we went over to his farm, sat on the stool and tried to imitate Grandpa's hand movements. We must have done it correctly as we soon had a half a pail of milk. We were very proud of ourselves until Uncle Martin deflated us by saying, 'Any fool can milk that cow. She ain't particular like your grandfather's.'[1]

Perceptive consumers have taken a strong interest in the implementation of safe, humane livestock handling. It is important for farm staff to develop an understanding of how animals learn and how to gently prompt intended animal responses. Because animal and human contact is often cited as the leading cause of injuries on the farm, the safety of workers and animals is a prerequisite of any humane handling education and training.[4,7,8] Even though livestock and human relationships vary the world over and every livestock operation is unique, the handling history of the cattle and the behavior of the handlers are surely different. Understanding the science discussed in previous chapters and applying it in humane cattle handling techniques reduces stress and injuries to cattle and humans.

Efficient Livestock Handling. http://dx.doi.org/10.1016/B978-0-12-418670-5.00006-8

The appropriate management of cattle behavior demands precise application of unambiguous language and technique. Anthropomorphisms and cattle handling concepts regarding cattle behavior, temperament, and techniques that are not consistent with state-of-the-art knowledge are being shown to be less efficient. The use of the word "fear" is an everyday example of anthropomorphism. Fear can only be known if an animal is capable of consciously experiencing this state; people obviously are, but the conscious experience of cattle is unknown.[6] Such terms as "happy cows," cow "comfort," understanding how cattle "think," cows with an "attitude," "fear memory," and interpretations of symptomatic behaviors are other examples of anthropomorphism. We can use conjecture about these things based on human feelings and experiences, but we will never truly know.[6] Unfortunately, the human response to a nonresponsive animal is to increase the stimulus, often to the point of pain. This has obvious negative welfare implications, especially if learned avoidance behaviors are misinterpreted as belligerence or if innate escape behaviors are confused with fearful behavior.

While the application of science to livestock handling has proven to be safe and efficient, there has also been a proliferation of misinformation and pseudoscience. Pseudoscience is presented as science but lacks supporting evidence and often relies on boasts or testimonials instead.[15] A claim that troubleshooting animal behavior is made easier if handlers understand how livestock think is a perfect example. While that statement is undeniably true, it is scientifically impossible. Knowing what livestock think requires that the animal first have introspective awareness and then be able to explain its thoughts.

If the science does not have practical, on-farm application, it is a foregone conclusion that the information will not be incorporated into on-farm programs. It is important, then, to bring together the science and the practical applications of humane dairy cattle handling.

LEARNING IN CATTLE

The cause and effect of human behavior on the quality of life for livestock is an expansive discipline in the contemporary, vibrant, evolving human and animal relationship. Livestock handling is a human endeavor involving the combined effects of facility design, animal biology, the past and present animal experiences, and human involvement. It is the last of these that will be discussed, noting that consistent methods from handler to handler and day-to-day are critical for producing predictable handling outcomes.

Handling live animals involves three-dimensional interactions played out in real time. It involves knowledge of how to reduce stimuli, when to stimulate, and at what angle and intensity to apply any stimulus. Practice teaches us how and when to act or react. Emphasis needs to be placed on "what to do" and "why it is done," rather than instruction on what not to do.

Practical applications reinforce the principle that animals learn by association. This means that every interaction between a handler and a cow has potential learning consequences. This can happen regardless of whether a lesson was intended or not.[9] A *stimulus* is any detectable change in an animal's environment, and a *response* is the behavior or physiological event that results from the change.[10] Cattle also have the ability to differentiate between significant and insignificant stimuli. After all, at any given moment, all creatures are influenced by many varieties of stimuli, and they usually react to the most intense and challenging one. A primary job for handlers is to be certain that handling stimuli are intended: then the outcomes will also be intended, efficient, safe, and humane.[14]

Inadvertently teaching cows to turn away from the parlor or teaching cows to run by handlers during sorting can be dangerous to humans and cattle. In late 1899, trail boss Frank Flood taught his cattle drovers to "never let a cow turn away from the destination."[16] That was true then and it is true in modern cattle production. Cattle that learn to rush past handlers will likely run past humans everywhere later. Running is potentially unsafe for all involved.

Trial-and-error learning is the most common form of learning for cattle. Handlers can make the best use of this type of learning by teaching cattle where to go using subtle movements. This also means that it is important to use consistent human body positions, postures, approaches, and timing. In practical terms, as a handler steps toward the cow and barely enters her flight distance, the animal responds by moving away to avoid the person. Her reward is that the human stimulus is reduced. Timing of the reinforcement must be closely associated with the desired outcome in order to provide meaning. The consistency of worker movement and the immediate reinforcement to the cow are the critical elements for cattle learning.

Over time, handler movement and cow movement toward the parlor will occur together in a fairly consistent way. However, it is important to recognize that while the handler's movement usually is associated with the cow's movement, it does not always cause the cow's response. In other words, even if a handler steps toward the cow, her movement toward the parlor may have already started for another reason, such as the radio coming on for afternoon milking. Either way, the behavior of the dairy cow becomes a predictable consequence of moving toward the parlor or the working chute.

While trial-and-error learning can be successfully used when associated with desirable reinforcement, avoidance can be learned too, particularly if the trial behavior is associated with pain or startle. This undesired learning is associated with locations, activities, and even specific people.[14]

EFFECTIVE VERSUS EFFICIENT HANDLING

Effective methods for handling animals are found everywhere livestock reside. Some effective methods were developed thousands of years ago, and some have been described by the modern day animal "whisperers." Many of the effective handling methods used today rely on hypothetical explanations of animal behavior extrapolated from descriptor terms used for human behavior.[2] While there are many customs or rituals that result in effective handling today, efficient handling in modern livestock production is intended to create valuable, productive behavior resulting in beneficial learning for the handlers and the animals. If the method is efficient, it is effective, but the reverse is not necessarily true.

The primary objective of effective handling is to get the job done irrespective of any consequences. Young animals raised in restricted space confinement have a tendency to run everywhere they go. This teaches animals to run every time they encounter human presence or unfamiliar places. For example, teaching animals to run to the milking parlor gives the appearance that the animals are "happy" to be milked. No handling system is perfect, but cows running toward the milking parlor will fill the parlor with adrenaline-saturated cows. The result is effective but it has consequences. The adrenaline suppresses milk letdown. While running is effective at getting the cows to the parlor, it is not efficient because harvesting almost all of the milk is the primary objective.

Effective handling is found everywhere, but efficient handling means more value for the effort. It gets the job done while avoiding many of the unintended consequences. Efficient handling relies on how animals can learn on their own. If naturally occurring escape behavior works for the heifer, then she likely learns that escape works and she will employ escape behaviors. Efficient handling walks cows toward the parlor with the least production of adrenaline and the optimum milk letdown. The efficient handler will either devise methods that prevent cattle from escaping or retrain escape-prone livestock. Retraining is almost always more difficult to establish than the original lesson, so the approach of teaching animals to herd under control rather than to escape out of control is much more efficient in the long run.

Pressure and Release

Efficient handling applies a stimulus based on need and not an "all-on or all-off" approach. Animals that are inadvertently taught to move a few steps and then stop, as happens with tail twitching or the use of the electric prod, is an example of what happens when a stimulus is applied and then stopped—like an on-off light switch. The handler may not have meant to teach the animal this unnecessary stopping, but he did. Applying a stimulus in this manner is referred to as the "pressure and release" method. The technique is to apply a pressure to the animal, causing it to react. The senses of the animal take in the stimulus, and the animal's instinctive or learned behaviors provide the reaction. "Stimulus at a distance" can be applied to untouched animals because their vision, hearing, and smell capabilities provide information to the cow about the distant human stimulus. If the human is close enough to directly or indirectly touch the animal, then "physical pressure" is an accurate term. Efficient techniques applied by physically touching cattle and selectively touching specific areas are as important as is the method and manner of approaching livestock.[10] Timing, warning, angle of approach for touching, specific contact location, and diffuse versus focal touching are critical elements in successful, safe physical contact with animals.

Release of pressure philosophy teaches that once the stimulus pressure begins the desired action, the stimulus is stopped—the "release." Just as with the light switch analogy, the release stops the human stimulus on the cow, and she typically slows or stops. Often, she then checks her surroundings. This creates a start-stop reaction in cattle rather than a smooth flow.

Pressure and Reduce

The "release" component of the "pressure and release" philosophy can be replaced by the more efficient "reduction" of a stimulus, another description of directing or managing animal behavior. Releasing the stimulus eliminates the ability to constantly manage the animal's behavior. Reducing the stimulus allows the handler to continuously manage the stimulus presence and intensity. The resulting animal reaction is managed and directed rather than corrected, and there is no interruption. This is efficient. In this manner, an animal's response to a stimulus is micromanaged using negative reinforcement. There is no need to correct behavior. Micromanagement with human actions is analogous to the rheostat control in a light's dimmer switch. The rheostat stimulus allows for uninterrupted or continuous animal stimulation and learning that is controllable and manageable.

Effective method proponents suggest that humans develop techniques for animal training because we purport to know what animals want, think, and feel. That is not currently possible and subsequently is not helpful either. Why worry about what cows think? It promotes guessing at what to do instead of developing handling strategies that are founded in what we know about animal behavior. Efficient handlers understand that animals can learn to distinguish between relevant and irrelevant stimuli. It is important to understand that we train animals from the perspective of their ability to learn. Handlers who concern themselves with what the cows are doing rather than what the animals are thinking are vastly more efficient at creating and managing animal flow in any facility.

BASIC PRINCIPLES OF HANDLING

Most animal welfare advocates agree that a low-energy low-stress philosophy is more humane than higher aversive energy tactics. Efficient livestock handling is best described as using low-energy approaches, efficient techniques, gentle tactics, and safe procedures. This promotes livestock well-being through the humane, judicious, efficient, and discriminating use of human stimuli. Appropriate handling empowers the human and animal relationship through the proper timing and intensity of stimuli.

The most important factor for producing efficient animal handling is to understand the difference between correcting an inappropriate behavior and directing an animal's behavior so that undesired responses do not occur. The stimulus is efficiently timed, and the outcome is intended instead of being late or requiring the reapplication of the stimulus. Efficient timing and intensity implies consistency of the human actions.

Efficient handling advocates for the understanding of how animals learn and respond to stimuli. Animal locomotion is limited to eight potential behaviors: forward, backward, left, right, up, down, some combination of the others, or nothing at all. The goal of efficient handling is to limit the potential for counterproductive responses through the use of quiet, unhurried, consistent, appropriately-timed human behavior to obtain optimum animal flow. This promotes predictable, unhurried animal behavior. Consistency and timing of low-energy human action is probably the most important livestock handling consideration.

The uniqueness of the dairy industry is that people interact with cows several times a day for multiple years. This represents numerous opportunities to create positive human and animal interactions. Modern dairy practice has become dependent on teaching people how dairy cows learn and respond to the behaviors of people in a dynamic environment. Every interaction between people and cows can shape the future behavior of both. These interactions can have positive, neutral, or negative outcomes. Humane, low-energy handling of livestock helps people become aware of the influence of human behavior on livestock responses, so on-farm education and training programs are a must. All livestock handlers on a farm need to exhibit similar behaviors when handling the animals. Inadvertent, unintentional, and subtle but conflicting human behaviors can cause cattle to become reactive.

Because cattle are gregarious and tend to learn through experiences, roughly 80% of dairy cattle on modern dairy farms are relatively easily handled and can be herded from pen to parlor and back without crisis (Figure 6.1). The remaining 20%, usually the young, distressed, or infirm, require more effort, care, and skill to be handled optimally. These require a disproportionately greater amount of time and effort to manage. It is important to handle this 20% group as efficiently as possible. It also presents an opportunity to discuss the difference between effective and efficient handling of reactive livestock.

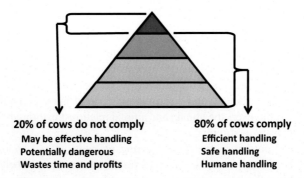

20% of cows do not comply	80% of cows comply
May be effective handling	**Efficient handling**
Potentially dangerous	**Safe handling**
Wastes time and profits	**Humane handling**

FIGURE 6.1   There are two populations of cows on a dairy farm relative to their ease of being handled. Most are relatively easy to work with, but the remainder waste a lot of time and can be dangerous. © *Don Höglund (2015).*

Regardless of farm size, there are several general age populations of cattle on a dairy. Each animal and each age group have different experiences so what they have learned varies. They also have differences in genetics. Both can be evidenced by the relative reactivity to any stimulus. Individual cattle that are highly motion sensitive are more reactive, so they try to avoid moving objects. Steers and heifers are reported to be more sensitive to motion than are older bulls or cows. This may reflect effects of sex, age, or experience. The cattle that are easily startled by sudden movements are the ones most likely to become agitated during handling.[5]

As is true for all animals, a dairy cow's response to a stimulus is dependent on sensory input. Although the specific limits of a cow's senses were discussed in Chapter 2, a few reminders that they are different than human senses are helpful.

The most important functional aspect of vision in cattle is the ability to identify an object as separate from its surroundings, as singled out by motion, brightness, texture, orientation, and color (to a small degree). The lateral position of the eyes and the ears on the skull of a dairy cow allows for a panoramic horizontal visual and auditory field, making it difficult to sneak up on her. The binocular visual field is less than that for a human, but it is used to make accurate judgments regarding the path and distances of a moving object. Evolutionarily, the visual capabilities developed to avoid predators by spotting the subtle differences between safe and not safe environmental events.[1]

The interaction between visual and auditory perception helps dairy cows localize the source of the sound.[3] Generally, the ear and the eye on the same side work in unison, except perhaps during aggressive behavior where the eyes are directed toward the pursuit and the ears are pinned or lay flat. Watching the ears can help determine the general direction the dairy cow is looking and subsequently whether attention is directed at the handler or at some other stimulus (Figure 6.2).

FIGURE 6.2 Ear position indicates direction of vision. Of the three heifers, the one on the right has her attention directed at the camera; the center heifer's ear position indicates that both eyes are looking behind her; the animal on the left is directing the vision of both eyes backward. The heifer in the back is looking back with the right eye and forward with her left eye. © *Don Höglund (2015).*

The sensitivity of cattle hearing makes loudness equivalent to hitting.[2] Because of this, talking or shouting is not particularly helpful in cow handling and can cause unintended consequences.

Dairy cattle initially tend to face a stimulus so that vision can provide a more accurate assessment of what might be happening. To take advantage of this, a person approaching the cow should come in a straight line path to lend predictability. A crooked or arching path toward the reactive animal requires more effort to maintain a focus on the person approaching, and the harder it is to keep this focus, the more likely the cow is to start moving away. The exact withdrawal distance and the extent of the response to stimulus can vary between cows. Even within the same individual, the withdrawal distance can vary depending on prior experiences, distances, boundaries, husbandry practices, age, other environmental stimuli, the health and well-being of the animal in general, and even the time of day.[16]

If a stimulus suddenly or intensely arises from behind, animals tend to startle or turn abruptly to capture a visual image to couple with the sound. Startling dairy animals is dangerous because someone could be kicked or run over as the startled animal tries to escape. Handlers should signal the animals if approaching them from the rear. A soft whistle or word is appropriate. Given the lack of other options, dairy cows will defend themselves when the person approaching reaches the critical distance. The response can vary from a defensive posture to overt aggressive behavior.

Past behavior of any animal is effectively the most reliable indicator of future behavior. Dairy cow handlers frequently refer to the flight and critical distances as arbitrary measures of how much encroachment an animal will endure prior to fleeing or fighting. The main limitation of judging animal behavior is that handlers likely do not know the living history of the animal, moment-to-moment or day-to-day. Since animals can learn from every interaction with the environment, what handlers observe is what handlers face during that moment, in real time. Handlers get what cattle behavior they get when they arrive at the pen. A key point is that handlers can induce animals to move, turn, rate, or stop by exerting and manipulating seamless stimuli.

FROM THE MATERNITY WARD THROUGH WEANING

The long-term economic health of the dairy industry depends on the well-being of the calves. In the two-year span from birth to first milk after parturition, the most important events to the development of dairy herd handling behavior happen in the first 3 months. By incorporating efficient handling into the everyday routine of young stock, producers can avoid losing 2 years of a training opportunity for both animals and staff. The initial human and animal interactions provide profound occasions to establish a low-energy relationship between all employees and the future members of the milking herd.

On dairy farms there are several populations of animals: near-term pregnant cows, freshly calved cows and heifers, calves, weanlings, juveniles, bred heifers, dry cows, bulls, and milking cows. Though 80% of all these animals are handled effectively, all could be trained to handle more efficiently. The most efficient way to categorize the various aged groups is to consider grouping them by virtue of their place in the overall life cycle of the farm.

Peripartum Cows

Peripartum is that period occurring during the last month of gestation (up close cows) or the first few weeks after delivery. From what we know now about epigenetics, low-energy handling of the cows is important to reduce stress on the cow and her calf, even if unborn. Handling during the near-term period is as important as it is at any other time in her life. Special care is of particular importance for first calf heifers. After all, the parlor is usually an unfamiliar place for first calf heifers, but it does not have to be. Handling can include familiarizing the heifer with the parlor or something that looks similar.

Dairy cows usually calve at night. This trait probably evolved to give the newborn time to get steady enough on its feet to escape, if necessary. For this reason, it is optimal to provide a spacious, well-bedded calving pen, such as a grasslands pasture. Low intensity lighting in the maternity room simulates nighttime and may help mimic the isolated area a free-roaming cow would seek.

Until the last few decades, dairy calves learned about life while following their mothers from pasture to pen to parlor. In modern North American dairies, this rarely happens anymore. Thus, every interaction between the calf and the handler has the potential to influence the future behavior of both. Maternity ward handling on the dairy farm should be considered the first step in teaching man and animal about each other. In the early hours after birth, the fewer stimuli, the better. Vigorous rubbing may be necessary to stimulate a calf's breathing, but handlers should otherwise move slowly and create as little noise as possible.

Most cows held in spacious birth pens will deliver healthy calves. Some producers choose to leave the calf with the cow for minutes to hours, and others place the newborn in a maternity ward soon after birth. Either way, handling is just as important for training the calf to accept humans as it is for any other time in the life of the animal. When calving assistance is needed, all efforts should be to extract the calf and place it in the maternity ward with the least amount of excitement as possible.

For the cows, herdsmen should remember that unfamiliar locations cause animals to stop and evaluate the situations. Calm, slow handling in unfamiliar surroundings will usually end with positive results. Rushing a cow into the colostrum chute and rough treatment of the udder will teach the heifer that humans, chutes, and confinement are associated with startle and usually pain. As with other animals, cows learn by association of events, including biological ones.

The Calves

Calves spend the first 45–70 days of life learning that food and water come from humans. As a result, heifer calves learn to face human handlers. At some point, the calves must learn to herd, or face away from handlers and move slowly from human presence toward a chute, some form of transport, or to and from a milking parlor. Heifer calves are usually handled in groups in much the same as we handle steers. Then one day about 2 years later, these heifers have their first calf and are expected to suddenly behave like experienced milking cows. People forget the importance of what cattle might have or have not learned already.

All stimuli are new to the calves, so familiarizing the calf slowly and gently to new sights, sounds, smells, tastes, and touches can go a long way in creating a calm addition to the future milking herd. Rough treatment will likely create reactive calves and reactive calves are hard to handle. Think slow and gentle. Pushing gently does not create the reactivity in young animals that pain or startle cause. Moving calves from place to place in the maternity ward and on into hutch life is a great place to begin introducing calves to react minimally to human and equipment presence (Figure 6.3).

The senses of calves are fairly well developed at birth so they can respond to environmental stimuli within hours. Because calves learn every waking hour and have no previous learning to interfere with new experiences, their tendency is to quickly learn to avoid rough or intense events. Aggressive handling teaches calves that humans are aggressive, and the resulting escape or fight behaviors are rewarded because the problem stops. This leads to calves that exhibit running or aggressive behavior when humans are present—a potential danger. When escaping, their attention is directed behind them, so calves could run into

FIGURE 6.3 Calves are particularly sensitive to stimuli because all things are new to them. From birth, gentle handling when toweling them off and careful moving (e.g., using a wheelbarrow instead of physically carrying them) are important. © *Don Höglund (2015)*.

objects like walls, equipment, other calves, or fences. It is far more productive for future handling if calves are taught to react calmly to people and new events.

Low-energy handling of calves should start at their birth. All human handling should be gentle, slow, and predictable. Touching the calf and helping frail newborns is necessary, but using slow, predictable movements will avoid startling the cow and will begin to teach the calf to react calmly to handlers. Mother cows do not drag their calves by the extremities (the ears, legs, or tail). They nudge their calf to its feet. Humans, too, should nudge calves, even rubbing them vigorously with dry towels rather than handling them roughly. The difference between vigorous rubbing and rough treatment can be explained by the difference between touch and pain receptors in the skin. Slapping the calf or excessively jostling it comes to be associated with distress.

How young calves are fed should take handling techniques into account. Colostrum feeding by tube has the potential to teach handler aversion, so it is important to avoid rough handling. Though science suggests that it is best to provide three liters of good-quality colostrum to the stomach of the calf within 30 min and a fourth within an hour, it needs to be done as gently as possible. When bottle feeding a calf, the handler should think of the mother as a model for how a calf nurses. Udders are shrouded in shadows, and the calf's head is turned up and into shadowy darkness, not up into blinding light. Minimally, it is best to cover the calf's eyes when bottle feeding (Figure 6.4). Reactive calves will generally calm down when their eyes are gently covered.

Calves that have been roughly handled overreact to human presence and learn to run everywhere they are moved. That encourages other herd mates to run with them—through alleys, into chutes, into fences, into each other, and away from the milking parlor. They are prone to injuring themselves and the staff. Moving slower is actually faster.

(a)   **(b)**

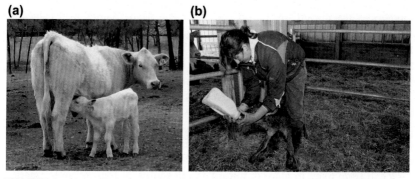

FIGURE 6.4   (a) A calf's eyes are normally shaded while it nurses from the cow. *(© Don Höglund (2015).)* When nursing from a bottle or pail, that is not true. (b) Covering or shielding a calf's eyes while it nurses simulates the shadows of the udder. *(Picture b is compliments of Hoard's Dairyman; used with permission.)* Reactive calves will often calm down if this is done.

FIGURE 6.5   (a) Dairy calves learn to face people because (b) they are fed from the front as calves. As heifers, they retain that behavior. © *Don Höglund and Jouni Pitkäranta,* www.cowhomes.com *(2014).*

Unlike beef calves, dairy calves rarely have an opportunity to learn from their mother and following her is the way a calf would normally learn to herd. She moves away from the handler, so the calf comes to associate moving away from handlers as the reward for human encroachment. Many dairies feed calves in hutches with bottles, buckets, or automated milking systems. That feeding protocol teaches calves to face handlers (Figure 6.5). When it is time for the fresh heifer to enter a milking parlor, she does what she has learned—face the handler. The need to teach calves to herd (drive) is common in most modern dairy operations (Figure 6.6). Calves raised with other calves in hutches tend to socialize better than when isolated in pens with solid walls.

Once calves are placed in a group, they develop social behaviors. Cross-sucking is a social behavior that can become an issue that needs immediate attention. Nearby human movement should be slow and quiet. Physically guiding a calf from place to place or toward a milk source, as long as neither startle nor pain is involved, is usually not harmful.

Herding is a lesson that should wait until calves are moved to weaning pens at 65–75 days of life. When this move occurs, it is best to allow 24–72 h for

FIGURE 6.6   Heifers need to be taught how to herd because of their tendency to face people. *Picture compliments of Hoard's Dairyman; used with permission.*

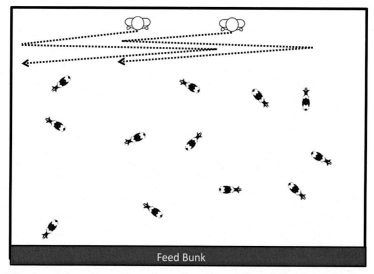

Feed Bunk

FIGURE 6.7   When teaching weanlings to herd, the handler enters the pen on the side opposite the feed bunk and first moves the calves toward the feed. If there is more than one handler, both move in unison. © *Don Höglund (2015).*

calves to learn the limits of the weaning pen before human interactions begin. Teaching calves to herd and "park" while they are small is a much safer process for everyone. Waiting until the animal weighs 1000 pounds before teaching it calm reactions to the human presence has obvious drawbacks.

Teaching calves to herd uses the same techniques as used on their older counterparts. It is best to enter the weaning pen on a side opposite from the feed bunk to avoid teaching calves to move off of feed every time they see humans entering the area. First, the calves are herded toward feed and held near the feed bunk for at least 2 min (Figure 6.7). A calf's attention span is approximately 10–30 s, so holding them for 2 min allows the handler to micromanage calf "parking" behavior. It teaches the animal to stay calm under command and stay where she was placed and held.

They are next herded to another spot in the pen (Figure 6.8). Continue teaching calves to herd by moving them from end to end of the weaning pen for at least three herding trials into different corners. Herd them to an end and hold them for at least 2 min. Then slowly and calmly herd them to the other end and hold them there for two more minutes. This takes 15 min for one person, each day for 3 days. After that, retrain that group for 2 or 3 days each time they flow into larger weanling groups or at least every 30 days or so.

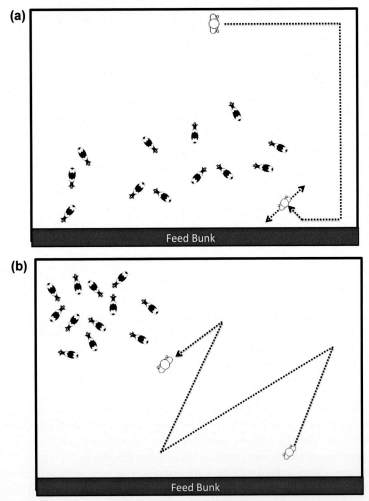

FIGURE 6.8 After the calves have been held near the feed bunk, (a) the handler walks along the edge of the pen to the corner opposite of where they are next to be herded and begins a slow side-to-side movement. (b) The handler then uses the zigzag pattern to move the calves to the opposite corner, where they are again held for a few minutes. © *Don Höglund (2015)*.

MOVING CATTLE

Where the person stands relative to the flight distance and the vision of the cow will determine which way her body will move. During movement, the cow's head position will turn in order for her to maintain visual contact with the handler. If the person is to the rear of the cow, she has a tendency to arc forward and away from the handler rather than move in a straight line. The handler's beginning position is of primary importance when moving cattle forward. As shown in Figure 6.9, if the cow has to turn her head to watch the approaching person, her body motion will tend to arc around him as she moves away. While the goal may have been a straight line movement, an arched one occurs instead. If instead the handler approaches directly from the side, the cow can see him but has the option of moving forward, sideways, or backward to get away (Figure 6.10). Another common mistake is shown in Figure 6.11(a). Here the handler starts near the cow's rear, walks parallel to her body, and then moves slightly toward her as he approaches the midsection. This will usually cause the cow to shift away and forward, with the forward motion caused by the initial rearward movement of the handler. A similar result occurs if the handler begins near the cow's shoulder and moves toward her hip (Figure 6.11(b)). Though moving forward is the intended direction and both of these are effective in gaining forward movement, they also condition the cow to move away from handlers instead of straight forward. It is more efficient to teach cattle to move as intended—straight forward.

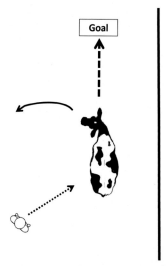

FIGURE 6.9 The cow's head and body are directed toward the approaching stimulus, causing the animal to arc to the left and away from the handler instead of moving straight ahead along the fence. © *Don Höglund (2015).*

FIGURE 6.10 When a cow is approached directly from the side, she has several options, depending on how close the fence is: move forward, move sideways away from the person, or move backward. The cow moves, but the handler has limited ability to control the outcome. © *Don Höglund (2015).*

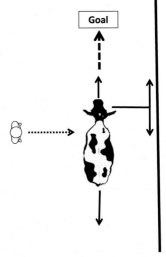

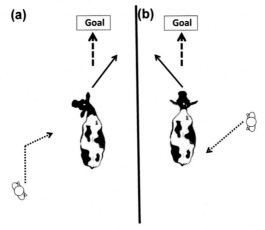

FIGURE 6.11 Effective forward movement is not necessarily efficient forward movement and it does not teach the cow to walk in a straight line. (a) When a handler walks from the rear parallel to the cow and then angles toward her slightly, she tends to move forward but away from the handler. This is forward movement but not straight forward movement. (b) When a cow is approached by a handler moving from her side toward her rear, she also tends to move forward but away. © *Don Höglund (2015).*

Moving parallel to the cow but toward the back half of her body is the most efficient way to move her straight forward. To use this technique, it is important to stay far enough to the side so that the cow can keep the handler in her vision and outside her flight distance so she does not have to turn her head (Figure 6.12(a)). This approach is more efficient because it allows the handler greater forward micromanagement of animal movement based on the speed of his movement. If the handler's position starts forward by the cow's shoulder

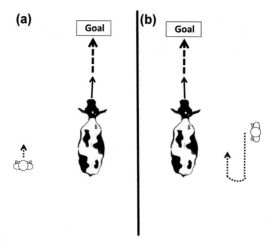

FIGURE 6.12 Efficient forward movement results when the cow moves straight forward. Walking parallel to her while positioned toward the back half of the body is useful to achieve the straight line. (a) When a handler walks from the rear parallel to the cow, it is important to stay within her line of sight. (b) When a cow is approached by a handler moving from her side, he should first move parallel to her toward her rear but outside her flight distance and then turn to move forward, parallel to her but toward her rear. © *Don Höglund (2015).*

The Herding Rule
(Frank Flood 1892)

Always toward the destination

Start movement – Keep movement

FIGURE 6.13   The herding rule of Frank Flood.

(Figure 6.12(b)), his initial movement is parallel to her but in the opposite direction. This will usually initiate cow movement in a forward direction. At the level of the flank, the handler angles forward again and moves forward again, remaining parallel to the cow. Working from the side of the animal, handlers can efficiently manage their presence, angle of encroachment or withdrawal, timing of movement, and rate of their action on the animal. Because cattle tend to keep at least one eye on humans, walking parallel to the cow tends to promote and maintain movement and prevent the starting-stopping problem (Figure 6.13).

In general terms, we think of cattle moving forward if the handler is positioned along the back half of the cow. The "point of balance," commonly mentioned in the literature as being at the withers, is really a subjective determination based on the influence of handler's presence and actions. The term also gives the impression that all cattle react in the same way. Reactive animals employ escape or fight behavior for what appears to be no detectable reason. These animals move when handlers approach from the front and from the rear as well. In the opposite perspective, nonreactive animals require a significant stimulus for initiation of

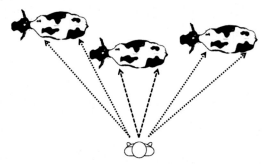

FIGURE 6.14 The starting position of the handler is different with respect to the three cows. If the handler moves toward the cows, the lead cow moves forward, the middle cow moves right and potentially forward following the lead cow, and the trailing cow likely moves backward. If the handler moves parallel to the animal movement, it will cause the animals to move forward, to slow, or to stop, respectively. © *Don Höglund (2015)*.

FIGURE 6.15 If the handler moves parallel to the animal movement, it will cause the animals to move forward, to slow, or to stop. © *Don Höglund (2015)*.

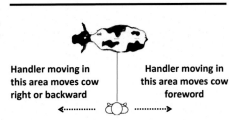

movement. It may even require touching, prodding, or an aversive stimulus. It is important to remember that humans tend to walk three to four miles per hour and cattle walk about two miles per hour. As a result, people can overtake dairy animals in only a few steps, and heifers will then either speed up or stop when the human reaches the proximity of the shoulder. Cows usually stop.

The usual objective is to move the cow(s) forward. However, efficient handlers can micromanage individual animal limb movement. This is done by incorporating a precise beginning position, angle of encroachment, and rate of handler movement (Figure 6.14). Handler motion parallel to and in the same direction as animal movement will soon cause the animal to slow and potentially stop, especially as the handler reaches the level of the cow's shoulder (Figure 6.15).

In reality, cattle can be moved from any handler position around the animal but not always as efficiently. Some just look back at the handler standing in the blind or area of limited vision. Handlers often try to herd animals from the rear, forgetting that dairy cattle have had no experience herding as calves and subsequently do not know what to do. Unless conditioned to herd, when handler stimulus enters areas of limited perception, such as behind the animal, the uninitiated animal will likely turn and visualize the handler. Animals that stop to look back or turn back in an arc are not moving forward efficiently.

On dairy farms, handlers interact with dairy cattle many times a day. Because of this frequent interaction, at some point their mere presence will have little or

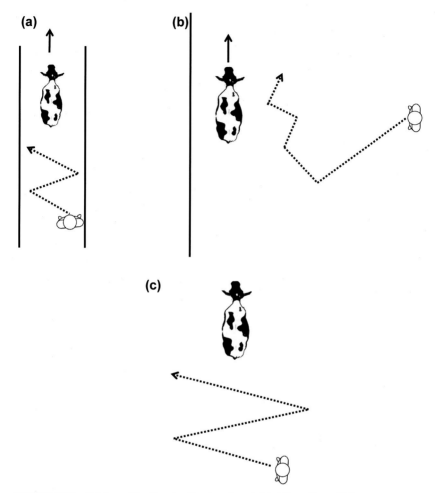

FIGURE 6.16 Driving cattle often must be done from behind them. To avoid the start-stop routine, the handler should use a zigzag pattern to remain within the cow's visual field as much as possible. (a) The zigzag pattern can be used in a chute or alleyway. (b) The zigzag pattern can be used while hazing a cow. (c) Herding cattle in an open field can also be done with a zigzag pattern. © *Don Höglund (2015).*

no effect on the animal. Training animals to move needs to be based on specific factors associated with a handler's presence. Things like a specific action, angle of movement, and rate of movement all combine to influence locomotion velocity and direction in livestock.

In an alleyway leading to a chute or transport, handler motion directly behind a cow in the blind zone will often cause the animal to attempt to observe the stimulus. If handler action is quick or loud, it can also cause the cow to startle. The way to prevent these things from happening is for the handler to move in a zigzag manner, crossing through the blind area calmly but quickly (Figure 6.16(a)). The goal is to

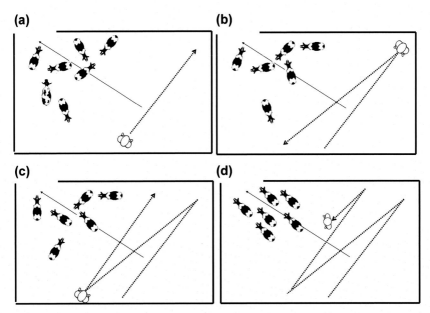

FIGURE 6.17   Using a zigzag pattern to move a herd of cattle. © *Don Höglund (2015).*

move from the cow's left eye vision to the right eye's vision and back, slowly, but predictably. This action stimulates the cow to move forward where there are only brief, noiseless moments where the cow cannot visualize the handler. It is recommended that handlers are calm, remain quiet, and keep their hands down. Hand and arm movement above the animal's eye level is a strong stimulus. If an increased stimulus is needed, hand movements should begin below the animal's eye level and rise gradually above as needed.

To move a cow along a fence line or a one-wall barrier, handlers should first move toward the animal's rear and then move with her but in a slight zigzag fashion, applying mild prompt and reducing stimuli while remaining in the cow's monocular visual field (Figure 6.16(b)). This action is predictable because the cow can see the person, and it serves to condition the animal to move in the intended direction.

In open spaces, such as pastures or large paddocks, the "eye to eye" zigzag method is again useful for herding individuals (Figure 6.16(c)). It is predictable and manages both sides of the animal. This long-used method allows the animal to see the handler most of the time.

When moving a herd, the same zigzag pattern is applied. The series of pictures in Figure 6.17 demonstrates how this can be efficiently done. The handler picks a target animal and walks in a line perpendicular to the desired direction of motion until he sees the cow's eye on the side toward his movement. The person then moves in a straight line in the opposite direction until he sees the other eye. The method can be used to move one animal, one animal in a group, or the entire herd simultaneously.

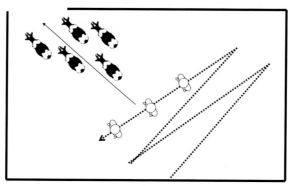

FIGURE 6.18   When more than one person is working the herd, all should be moving in the same direction and at the same speed so as to not give the cattle confusing signals. © *Don Höglund (2015).*

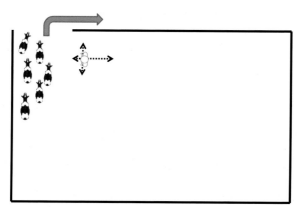

FIGURE 6.19   When cattle resist exiting a gate, they can be moved to the fence line and then stimulated to exit by using a left-to-right rocking motion. The handler also uses a forward-backward motion to add or relieve pressure relative to the flight distance. © *Don Höglund (2015).*

If more than one handler is present, all handlers move simultaneously in similar directions as if a string was connecting each handler to the others (Figure 6.18). This allows for a consistent handler presence and action, avoiding conflicting handler action. Once the gate is approached, the cattle make decisions about whether to maintain a fluid, continuous movement through the gate or whether to put up resistance. If the animals continue forward, the handler can continue zigzag stimulus. If, however, the animals resist exiting the gate, a different strategy is initiated. First move them to the far fence line by having the handler move to the side of the herd opposite the fence (Figure 6.19). Then he uses a left to right rocking-in-place motion to stimulate cattle to exit the gate. If two handlers are available, position one near the gate for the cattle to arc around and the second person is to drive the animals from behind and slightly to the side of the herd (Figure 6.20). If the handler is on the outside of the curve of

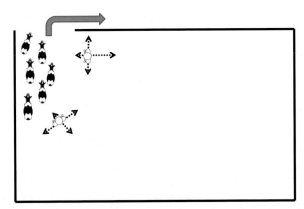

FIGURE 6.20   When two people are moving the cattle through a gate, one takes a position near the gate so the cattle will arc around them. The second person uses the back-and-forth zigzag motion to encourage continued forward motion. © *Don Höglund (2015).*

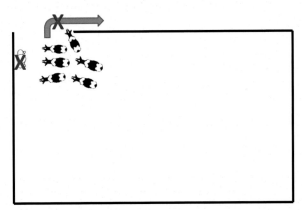

FIGURE 6.21   A handler should avoid standing on the outside of a curve the cattle are expected to make when exiting a gate. Cattle tend to investigate and discontinue the smooth forward motion. © *Don Höglund (2015).*

animal movement, as is shown in Figure 6.21, the cattle have a tendency to turn and look at the human, disrupting the flow and efficiency.

Stragglers and cows that break from the herd should be ignored because most of the time they return freely to the moving herd. Acknowledge it but maintain the integrity of the herd if going through a gate. Once the herd is through, the lobo can be redirected to the gate. If stragglers are disruptive, they need to be managed by penning them separately or returning them to the herd. Animals that are chronic problems should be worked in a separate pen until they respond appropriately to handler direction. Some cows learned to break away when they were unsuccessfully chased. The cow learned that breaking away was rewarding and that behavior continues. For some of these problem

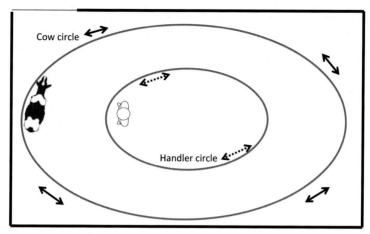

FIGURE 6.22 As a reactive cow circles the holding pen, the handler circles with her, stopping when she stops and moving when she moves. © *Don Höglund (2015).*

cows, it may be necessary to first circle them in a pen or paddock to remove the excess energy from the animal. Place the lobo in a pen by herself or, if too fractious when isolated, with some docile animals and let her explore the limits of the pen. The handler should continuously follow the problem animal regardless of where in the pen she goes, maintaining a relatively constant distance from her. Refrain from running or chasing. Gradually, the random wanderings become circular around the pen. The animal has searched the pen and has found no space that made the handler any less present. The handler should continue to follow the animal in an arc somewhat smaller than that made by the animal (Figure 6.22). The handler is to remain as quiet as possible so that the only handler-related stimulus is in the form of small circles. When the animal slows, the handler should slow. The moment the cow slows or enters a corner, the handler should stop. If it remains stopped, the person should back away a step or two (negative reinforcement). This lesson teaches the cow that the reward is found when she stops, and it starts to lower the energy in the pen.

The cow will move again and the handler is to follow her until she stops. As this lesson is repeated, the cow will gradually stand still longer when the handler withdraws a step or two. Reapply the backward motion each time the animal begins to move so she will stop again. The objective is to be able to park the animal in a corner and then follow it at a distance everywhere it goes in the pen. This method helps teach the cow to accept human presence and action while learning to herd toward some destination predetermined by the handler. The cow begins to associate her walking with a handler walking behind: herding has begun.

Once the handler has parked the animal, facing away and visualizing the handler with one eye, the handler should step forward on that side, slowly beginning encroachment. The animal moves and the handler can manage this

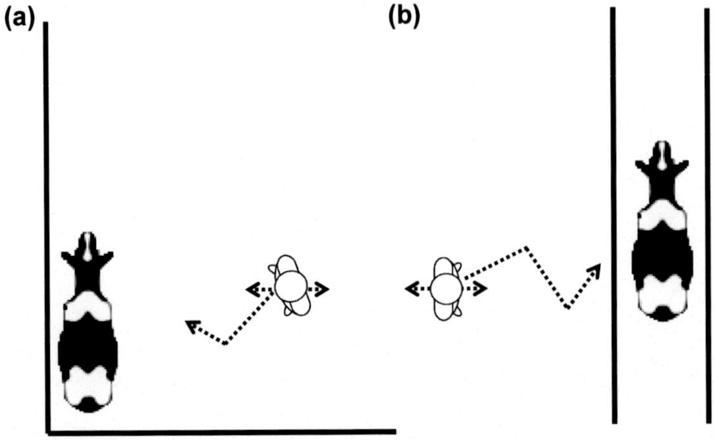

FIGURE 6.23 Reactive cattle can be habituated to human encroachment by slowly approaching them at a 45° angle until they hold still, at which point the human steps back and stops. As the cow becomes more comfortable with the presence of humans and stands still longer, a back-and-forth rocking motion is added. This can be done in either (a) an open pen or (b) a chute. © *Don Höglund (2015)*.

encroachment with slight movements, always ready to step back in order to reduce the stimulus on the cow. The animal and handler objective should be to walk, with the animal being herded slowly.

For highly reactive dairy cattle in an open pen or in a chute, the handler should approach at a 45° angle and stop when the cow becomes reactive, holding the position as long as animal is reactive (Figure 6.23). When the cow stops her reaction, the handler is to take one step back and hold, ideally for about 30 s. Repeat this several times. Eventually add in a forward-and-back rocking motion to move one cow foot at a time. This process habituates the animal to human encroachment and prepares her for additional herding training.

Efficient techniques may include touching the animal, particularly cattle that are used to interacting with people. How this is done is as important as is the method and manner of approaching her. Timing, warning, angle of approach for touching, specific contact location, and diffuse verses focal touching are critical elements in successful, safe physical contact. Think safety while being sure the cow recognizes someone is coming toward her. A startled animal is dangerous.

REDIRECTION PENS

The redirection pen is one method cattle handlers can use to manage livestock reaction to human presence and to move the animals to new locations containing unfamiliar equipment or transport vehicles. Bud Williams is credited with popularizing its use, thus the other name—the "Bud Box."

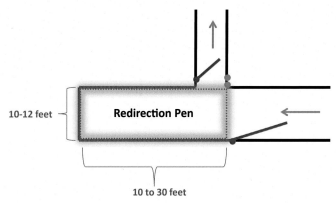

FIGURE 6.24 The design of a Redirection Pen (also known as a "Bud Box") makes it easier to direct cattle toward unfamiliar areas. The pen is 10–12 feet wide, 10–30 feet long, and has an exit chute that is the width of a single animal. The location of the gate hinges (●) and latches (●) and the direction the gates open are very important. © *Don Höglund (2015).*

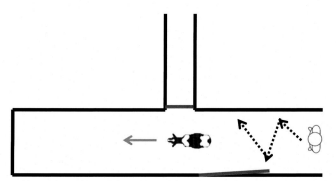

FIGURE 6.25 The cow or group of cows is slowly moved into the redirection pen and moved toward the far end. Note that the exit gate is closed. © *Don Höglund (2015).*

Cattle have a strong tendency to return to familiar housing or pasture, so handlers have been able to configure pens to take advantage of this innate behavior. On dairies, 10–12-foot long, 5–6-foot high portable panels are useful for creating a rectangle that measures approximately 10–12 feet by 10–30 feet (Figure 6.24). Where the gates hinge and how they swing are important. With the entrance gate open and the exit chute gate closed, cattle are herded toward the redirection pen in numbers appropriate to fill a transport compartment or the alleyway to a chute (Figure 6.25). After the cows have entered the box and moved slowly toward the far end, the entrance gate is closed and the exit gate is opened. The handler takes a position inside the redirection box just past, but near, the hinge for the exit gate (Figure 6.26).

In order to move cattle to the exit gate, the handler begins a side-to-side rocking motion (Figure 6.27). The amount of rocking and the forward movement of the handler control the amount of the stimulus, much like a rheostat

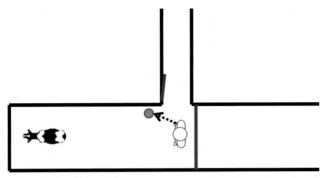

FIGURE 6.26   Once the cow or group of cattle has been moved into the redirection pen, the entrance gate is closed and securely latched. The handler fully opens the exit gate and moves to a position near the hinges of the exit gate. © *Don Höglund (2015).*

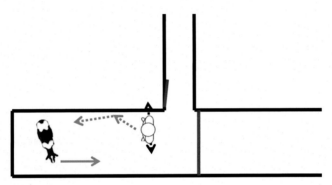

FIGURE 6.27   The handler begins a side to side rocking motion (◀····▶) to start the cow's motion back toward the gates. If additional stimulus pressure is needed, the handler might have to move slightly toward the cow too (◀···▶). © *Don Höglund (2015).*

does for lighting systems. Livestock tend to return to what is familiar and will return to the spot where they entered the pen. That is why the latch of the redirection pen will be most efficiently placed if it is juxtaposed to the entrance of the chute or transport alley. The handler walks parallel to, but in the opposite direction of the cattle (Figure 6.28). As the cow approaches the pen entrance point, she will see the exit to the chute and will usually enter the chute because of the tendency to curve around the person. Similarly, a group of cattle is worked the same way (Figure 6.29). As the last cow moves toward the exit, the handler follows her on the side rather than continuing toward the back of the box.

Training cattle to go through a redirection pen while young will familiarize them with this equipment early. This results in increased efficiency when they encounter this again because it is no longer a new or novel experience.

If the handler positions himself opposite the concave curve of cow movement, the reactive animal will attempt to keep at least one eye on the handler,

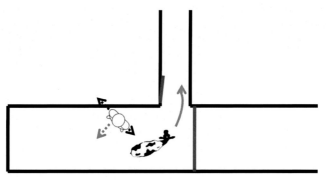

FIGURE 6.28   As the handler continues the rocking motion, the cow starts toward the area where it originally entered the pen, but will curve around the handler into the open exit area instead. As the cow begins to enter the exit gate, the handler walks parallel to her but in the opposite direction (◀···▶). © *Don Höglund (2015).*

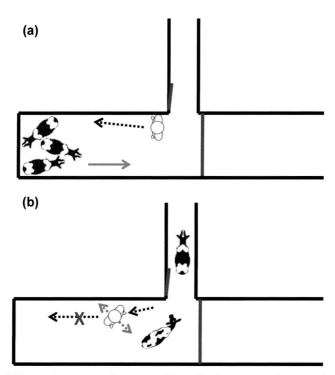

FIGURE 6.29   Moving a group of cattle out of the redirection pen is similar to moving a single cow. (a) The handler first moves down the side of the pen where the exit gate is located to move the cattle toward the entrance. (b) As the last cow approaches the entrance, the handler stops moving toward the rear of the pen as he gets even with the middle of the animal. A gentle forward-backward motion encourages the animal to move toward the exit. © *Don Höglund (2015).*

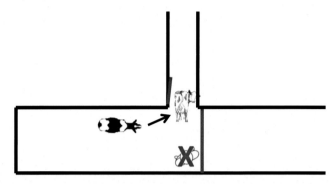

FIGURE 6.30 Handler position is important when using a redirection pen because the animals will arch around the person. Avoid standing by the latch or hinge of the entrance gate when cows are expected to enter the exit alley. Standing by the latch stops the cow from entering in the first place, and the second location, while farther away, causes cows to stop and look, which breaks the smooth, continuous flow of movement. © *Don Höglund (2015).*

causing the animal to try and look back (Figure 6.30). The flow of animal movement will slow or stop, defeating the purpose of timely, efficient, continuous movement of animals from their origin to the final destination. Cattle that learn to stop and look back in the chute may learn by association to enter chutes or alleyways and look back, creating stop and go animal traffic. The essence of good animal flow is to create movement and maintain it.

The redirection pen has modern variations, such as the sorting pie (Figure 6.31). Curved pens, though no better at helping inefficient handlers move cattle smoothly, are an aid for those handlers who understand that the animal will arc around the handler in a bid to keep at least one eye on the stimulus of the handlers.

Cattle may also stop if they associate a negative experience with a certain location, such as a gate opening, a pen, an alleyway, a chute, and a parlor. A causal relationship between resistance to locomotion and the precipitators for the resistance should be identified and the animal retrained. As a matter of efficiency, retraining animals is often more cumbersome than conditioning them properly in the first place.

DOWN AND IMMOBILE COWS

The humane handling of down cows is a difficult challenge because of the animal's size. Also complicating this is the long list of things than can cause a cow to go down—some needing more urgent care and others needing time. When faced with a down cow, handlers can improve outcomes by observing what animal is involved, what behavior is observed and its duration. Determining what should be done will be based on the conclusions of the first two observations.

The word "downer" has been co-opted to designate cattle in prolonged recumbency that are injured, feeble, or medically ill as the result of traumatic,

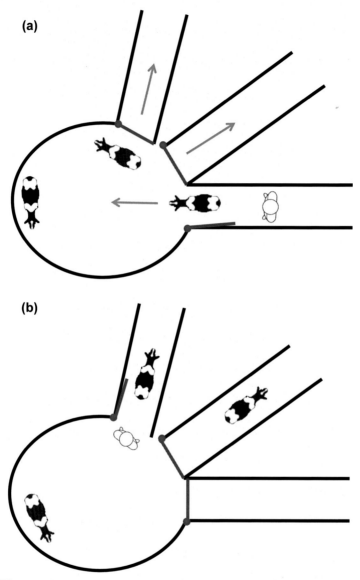

FIGURE 6.31   The sorting pie is a variation of the redirection box. It contains (a) an entrance and (b) multiple exit points leading to loading ramps, chutes, scales, or other pens. Note that there is a slight break in the wall of the sorting pie that serves as a walk-through space for handlers. © *Don Höglund (2015).*

toxic, metabolic, infectious, degenerative, and behavioral disorders. Cows in lateral recumbency need to be shifted into sternal recumbency as soon as possible and the cause of the dilemma resolved expediently, if feasible (Figure 6.32). Being recumbent for a prolonged time results in complications for the cow and

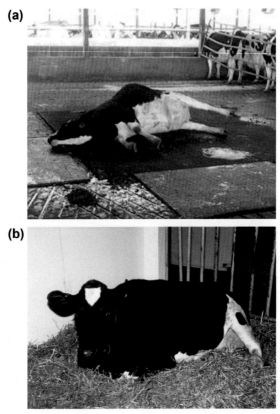

FIGURE 6.32   (a) Downer cows can be positioned in lateral recumbency. (b) Preferentially, they are in sternal recumbency with their legs pointed to the left side of their body. *Pictures compliments of Geof Smith; used with permission.*

reduces the likelihood of her getting up again. Pressure results in fluid being trapped in tissue compartments—a syndrome where venous flow is hindered due to animal weight on muscles yet arterial flow remains functional. The fluid pressure causes necrosis to connective tissue and nerves. This makes finding the cause of recumbency important (Table 6.1). Irrespective of immediate physical or medical support, cows that are not standing under their own initiative within 24 h also have a low overall recovery rate.[16]

Often overlooked in cattle handling is that cows that are alert, though down, or those that are standing but resistant to movement often receive aversive stimuli by handlers who mistake a physical problem from a temperament problem—an "attitude." Cattle avoiding locomotion and downers that cannot move endure aversive procedures in order to "motivate" them. Irrespective of intense punishment and pain applied to stimulate them to rise or move, they resist and even give up trying to respond.

TABLE 6.1 Types of Conditions Associated with Downer Cows and Cows That Are Reluctant to Move

Employ a Rapid Response System for Finding and Responsibly Resolving Down or Helpless Animals	Metabolic, Medical, or Issues Related to Cancer, Progressive Degeneration, Starvation, Toxins	Physical Injury Fractures Birthing Injuries Traumatic Injuries	Learned Helplessness Behavior (Nothing works for the cow)	Other Avoidance Behaviors
Down or standing dilemma	Covered extensively in literature	Covered extensively in literature	Misunderstood as "attitude"	Misdiagnosed as attitude
	CA^{++}/Mg^{++}, disease, toxins	Nerve/bone/malaise	Pain/learned avoidance	Learned or fight behavior
Solution	Diagnose/treat	Evaluate and support	Stop	Retrain
Water tank/bed	Sternal standing	Sternal ± standing	Always standing	Always standing
Primary need	Standing quickly <4h	Evaluate prognosis	Find relief retraining	Retraining
Movement Disposition Humane solutions	4+ hours use sling, sled, propped-up, water bed, tractor or bucket used humanely,	Sling, tractor bucket used humanely, sled, water bed but not water tank	Let the animal return to housing, milk there, and take to a training pen and retrain	Retrain the animal to escape versus to fight, then herd at a walk instead of avoidance or escape
	Watch for mastitis	Humane solutions	Solvable	Solvable

Inhumane resolutions to these dilemmas are highly visible welfare issues that negatively impact animal welfare, farm staff self-awareness, consumer perception, farm reputation, and industry image. The visibility of frustrated handlers effecting inhumane treatment effects consumer perceptions of all food-animal production practices.

Handling Recumbent Cattle

Laterally recumbent cows can regurgitate and aspirate stomach contents into the lungs. The animal should be manually rolled into sternal recumbency, preferably with her legs pointing out from the left side of the body (Figure 6.32(b)). The cow is supported in this position with straw bales placed under her side and shoulder or by a sling. She needs to lie in deep sand or thick straw bedding. Concrete surfaces without clean, substantial bedding should be avoided.

Commercially made water tanks are now available where the buoyancy helps suspend a downer cow. The water level needs to be monitored so that the cow can easily keep her head above the water level. It is also important that the water be kept clean and heated if the animal appears to shiver. Cattle that can stand for several hours with the aid of the water's buoyancy can then be encouraged to walk slowly from the tank on a nonslip surface to a well-padded area. Cattle that stand on all four limbs and those that are able to walk out of the tank after the first flotation treatment are more likely to survive. Reported success rates range from 37% to 46% in returning recumbent cattle to normal ambulation using water support.

Each day that a cow is down, handlers should attempt to gently help her stand. From sternal recumbency, the cow is rocked by gently pushing variously on her side as needed, up to but not including the point of using pain. Handlers can stand with feet pressed under the cow at a point below the cow's shoulder joint. If the animal struggles to rise, an assistant may help by grasping the very base of the tail and gently supporting the cow. Lifting on any other part of the tail can cause injury and should be avoided.

Slings that support the entire thorax and abdomen are useful if safely applied (Figure 6.33), but their use requires experience and a gentle approach. The raised position provides clinicians with an opportunity to auscultate, investigate, and perform whole body examinations on otherwise down cows. Hip clamps are used to lift downer cows, but the humaneness is still under study.

Animals that are not downers can learn to respond, but they can also learn that their responses are ineffective because something aversive will happen regardless of what they do. As an example, a cow that refuses to move along in a chute alley is subjected to electric prod pain, and she steps forward. Just as she begins to step, an overly eager handler shocks her "for good measure." For the

FIGURE 6.33 Slings can be used to support downer cows to help minimize tissue damage done when they are down for long periods. © *Don Höglund (2015)*.

cow, not moving and moving are subjected to the same aversive stimulus. So she quits doing anything—*learned helplessness*.[11]

The uncontrollability of the aversive stimulus created an animal that stands helpless and does not move regardless of the repeated or prolonged shocking. The cow has learned that she is going to get unpredictably shocked. The cow may even lie down and vocalize. Handlers often equate the "sullying," "balking," or "locking-up" behavior with an "attitude" in need of drastic measures. The animal is willing to endure the pain of repeated shocking or striking by a handler because, in reality, nothing she does escapes the aversive stimulus. Often, the negative situation also comes to be associated with the place it happens.

Prevention of learned helplessness and situational avoidance is always better than having to retrain an animal. Reapplying the pain or startle certainly is not going to resolve the issue. Since electric prods are rarely used correctly, they are best relegated to museums, not farms and ranches. Retraining the animal begins somewhere besides the location where the adverse event occurred. Because cattle are herd animals, the affected cow should be worked within a

FIGURE 6.34   Balking is usually associated with inexperience, human impatience, and previous adverse stimuli. © *Don Höglund (2015).*

group using the techniques of zigzag movements (as previously discussed). She will also need extra time to investigate new situations, such as entering a new pen or loading into trailers because she has learned that humans are to be avoided (Figure 6.34).

Moving Recumbent Cows

Moving a down cow to soft bedding in a good location will greatly enhance a healthy outcome. A method for moving the animal should be available and functional, and handlers should be trained and retrained in the efficient use of the device. Desirable locations include a roofed structure or wall protecting the animal from harsh sun, temperature extremes, or distressing weather.

Actually moving down cows is not simple. They are heavy and their size and limbs make lifting and moving them challenging. The primary concept is to do it humanely. That means not inducing additional pain, pulling on them, or choking them. Dairy managers should have the necessary equipment and training for dealing with the inevitable down or helpless cow. Sleds and slings are commonly used. Covered gates, plywood sheets, and doors can be useful if rolled over logs or pulled like a sled. Mobile water tanks, when empty, are helpful too (Figure 6.35).

Humans tend to attribute a cow's behavior to her fearing harm, rather than accepting it as a natural reaction to what is happening. By realizing that cattle are reacting to our actions, it is possible to deliberately change the way herdsmen and herdswomen handle cattle. When the handler can evaluate where the cow is and what she is doing, he can provide the efficient type and amount of stimulus to manage her behavior. This is effective and efficient dairy cattle handling.

(a)

(b)

FIGURE 6.35 Commercial water tanks can also be used (a) to move a downer animal from one location to another, as well as (b) to help position her in a vertical orientation. *Pictures compliments of Aqua Cow and Sandy Ingraham; used with permission.*

REFERENCES

1. Gilger BC. *Equine ophthalmology*. New York: Elsevier; 2010. p. 536.
2. Gosling S. From mice to men: what can we learn about personality from animal research. *Psychol Bull* 2001;**127**(1):45–86.
3. Heffner RS, Heffner HE. Visual factors in sound localization in mammals. *J Comp Neurol* March 15, 1992;**317**(3):219–32.
4. Langley RL, Morrow WE. Livestock handling—minimizing worker injuries. *J Agromed* July 2010;**15**(3):226–35.

5. Lanier JL, Grandin T, Green RD, Avery D, McGee K. The relationship between reaction to sudden, intermittent movements and sounds and temperament. *J Anim Sci* June 2000;**78**(6): 1467–74.

6. LeDoux JE. Coming to terms with fear. *Proc Natl Acad Sci* February 25, 2014;**111**(8):2871–8.

7. Lindahl C, Lundqvist P, Norberg AL. Swedish dairy farmers' perceptions of animal-related injuries. *J Agromed* 2012;**17**(4):364–76.

8. Merlo C. Dairy's own animal-care activists. *Dairy Today* January 2015;**30**(1):8–10.

9. McGreevy P. *Equine behavior: a guide for veterinarians and equine scientists*. 2nd ed. New York: Saunders; 2012. p. 378.

10. McGreevy P. *Equine behavior: a guide for veterinarians and equine scientists*. New York: Saunders; 2004. p. 369.

11. Overmier JB, Seligman MEP. Effects of inescapable shock upon subsequent escape and avoidance responding. *J Comp Physiol Psychol* February 1967;**63**(1):28–33.

12. Pajor EA, Rushen J, de Passillé AMB. Aversion learning techniques to evaluate dairy cattle handling practices. *Appl Anim Behav Sci* September 2000;**69**(2):89–102.

13. Rath S. *About cows*. Stillwater (MN): Voyageur Press; 1987. p. 221.

14. Rushen J, Taylor AA, de Passillé AM. Domestic animals' fear of humans and its effect of their welfare. *Appl Anim Behav Sci* December 1999;**65**(3):285–303.

15. Shermer M. *Why people believe weird things: pseudoscience, superstition, and other confusions of our time*. New York: Freeman and Co.; 1997. p. 306.

16. Smith B. *Moving 'Em: a guide to low stress animal handling*. Kamuela (HI): The Graziers Hui; 1998. p. 352.

Chapter 7

Beef Cattle Handling: Practical Applications of Science

To country people Cows are mild,
And flee from any stick they throw;
But I'm a timid town bred child,
And all the cattle seem to know.

T.S. Eliot

Obvious trivialities started more stampedes than anything else except storms: A stray dog sneaking up and selling around a sleeping animal on the edge of the herd; a bunch of wild hogs rooting into the bed grounds; the sight of a haystack after dark in a field into which a herd of old and leery beeves had been turned for the night 'so that the boys could get some rest'; the cough of a cow; a human sneeze; the snapping of a twig; the sinking of a circling horse's foot in a prairie dog hole. It was the suddenness of a sound or movement rather than unfamiliarity that made the drags wake up and forget all about sore feet. A polecat would come sashaying along in its nonchalant way in the edge of a herd; some steer, awake and investigative, would begin following the hopping creature. Then, noticing the approaching monster, the little hair-ball would stop, pat its forefeet against the ground and go to jerking its tail with comical swiftness. Mr. Steer had never seen this performance before—and it was 'so sudden.' He would wheel and snort, having no intention of starting a stampede, but a moment later he was part of the panic terror he had caused.[2]

There is an art to working with beef cattle, and as with any art, it can be improved with thought and practice. Handling talent is born out of the application of applied behavioral science, while the development of efficient techniques, hand-eye coordination, and expertise evolve through consistent repetition, timing, sequence, and the controlled mechanics of human actions. Inconsistency in starting positions, the timing or handler actions, and the rate of movement by a handler from interaction to interaction are the enemies of efficient and safe animal handling. Talent in any animal handling endeavor cannot be taught; it must be earned. Handling skill comes from lots of practice and from incorporating what science has shown us to work. And it should always be humane.

Efficient Livestock Handling. http://dx.doi.org/10.1016/B978-0-12-418670-5.00007-X

Ensuring beef cattle welfare is the responsibility of those who work with them because domestication made them dependent on humans. Behavioral responses are among the most pertinent indicators of animal well-being,[7,12] and they are particularly useful for safe, efficient cattle handling.

THE MIND IS UNKNOWABLE

One of the first lessons in livestock handling is to quit trying to explain the animal's behavior in terms of what it might be thinking or feeling. Suggesting what an animal is thinking is merely guessing at it and constitutes a prime example of the "illusion of knowledge."[1,13] Instead, we are dependent on observing what it is doing. Ambiguous, subjective labels applied to beef cattle behavior, such as "unmanageable," "unruly," "uncontrollable," and "afraid" are hypothetical explanations of behavior. Well-being concerns for both livestock and humans arise when handlers assume that livestock are willing, benevolent, or malevolent participants in search of human leadership.[10] A handler cannot know that the steer appreciates going somewhere, trusts the handler, is willing to cooperate, is friendly or afraid, and moves effectively because it was her idea.

Irrespective of the mental motivations of a bovine, each individual is part of a species that evolved by being able to detect challenges, find nutrition, remain hydrated, thermoregulate, reproduce, and fight or flee potential danger.[6] The development of unwanted behaviors has welfare implications for beef cattle because they are mislabeled as "belligerent" and deserving of aversive treatment.[9] Punishing cattle is wrong. Assessing what in the environment might have prompted the behavior is appropriate. Science-based handling starts by relying on what animals are doing instead of what they are hypothetically feeling or thinking.

Livestock are not likely to know the human by his intentions. They observe our behavior and respond according to the stimulus we present. While observational learning may or may not occur in cattle, handlers can improve handling skills by studying beef cattle behavior.[8,14] Observations allow us to understand the subtleties of how cattle learn: reliable science-based sources add perspective. An efficient animal conditioning technique is the means by which a predictable relationship with the farm animal can be developed. Whether in a pasture, breezeway, free stall, capture chute, or corral, cattle can be conditioned to stand calmly and quietly within a few consistent repetitions.

Rather than describing the attitude of a range-raised steer escaping or avoiding an open gate as "belligerent" or "fearful," handlers could simply ready themselves for the steer to exhibit this behavior and watch for circumstances that predict or trigger it. If behavioral signs are observed, a handler can intervene to reduce or prevent the unwanted behavior from happening. For instance, by ensuring that the steer always faces the destination, like an opening, a handler can be prepared to prompt the steer to move forward just before it stops or reverses course. If the handler waits for the steer to stop or turn before renewing the prompt for forward locomotion, stopping and avoiding the gate can become part of the steer's learned pattern. The same is true when working with herds instead of individuals. The first escape or avoidance attempt is a trial, the second is a repeat of the first and is likely a reinforcer. The second and third attempts help condition the steers to avoid gates as a learned behavior. Efficient handlers are prepared to stimulate the herd from a safe and efficient position, angle, rate, and intensity. They are also prepared to immediately reduce the stimulus at the appropriate time.

Senses Guide Behavior

An animal reacts to the world around it based on its sensory input, inherited tendencies, and previous experience. These are different in prey species than they are in humans, but understanding the differences is helpful for efficient handling. Humans are very dependent on vision, particularly binocular vision. Cattle are more dependent on a wide monocular visual field to scout for predators. The horizontal pupil opening is also important for this wide field. Depth perception is sacrificed, but that is not particularly important to cattle anyway.[3] The direction the ears are pointing helps handlers determine what a beef cow may be looking at because the ear and eye on the same side generally work in unison.

While the audiogram for cattle is very similar to that for humans, there is one important difference. Cattle can hear softer sounds than people can. Loud noise is stressful to them. It has been shown that shouting and loud noises cause the same physiological response as hitting does.[11]

The senses of vision and hearing are arguably the two most important senses involved in escape, avoidance, or aggressive behaviors. If there is plenty of time, avoidance usually happens. Since cattle are not able to move particularly fast or go very far at a fast speed, they are well prepared to fight. Given a lack of other options, beef cows will defend themselves, progressing from a defensive posture to overt aggressive behavior.

Reactive distances, as described in Chapter 2, are particularly relevant to beef cattle. There are different animal responses to handler stimuli depending on the distance from the animal and the intensity of the stimulus (Figure 7.1). Those cattle with minimal human contact have a flight distance for an approaching person that is farther from them than it would be if they saw people every day. For show steers, the flight distance would be reduced to zero and their reaction would be dependent on touch sensitivity instead. Less handled cattle would act aggressively when the person reached the critical distance. With tame cattle, there is usually an increase in reactivity when a visual stimulus moves to within 20 inches (0.5 m) of the animal's eyes—the limit of their ability to focus well. Handlers are advised to slow the encroachment and even withdraw a few steps to allow the cow to regain focus on the handler. The other option is to stop and wait for the cow's reactivity to abate. Vocal or hand prompts are a method to

FIGURE 7.1   The cattle stopped are naturally curious but will stop approaching something new when they reach the edge of their flight zone. The direction the ears are pointing is indicative of the direction the cattle are looking. © *Don Höglund (2012).*

condition the animal to predict that handlers are about to touch them. Wiggling a hand and then touching the animal with the back of the hand is one excellent method for establishing a pattern during the motion to touch the animal.

FROM PASTURE TO PEN TO CONFINEMENT

While events such as cutting, roping, and working cows may be fun to watch or participate in, they have little resemblance to the humane handling objectives needed for efficient cattle management. During these events, cattle are running. It works for the contest, but not on the ranch. There are limited sorting needs and emergency situations where running might be necessary, but unless a farm or ranch lacks chutes, sorting pens, and traps, chasing and roping are not advised today.

Beef cows learn primarily by trial and error. Beginning handler position, timing of stimulus, consistency of motion, and immediate reward or reduction of stimulus are the critical elements in developing adaptive learning. Food is a strong reinforcer. So is the prompt relief from aversive stimuli, mainly by negative reinforcement. For example, if a human moves toward a cow and she moves away, the averseness of human encroachment is relieved. By whatever means, the removal or reduction of a stimulus reinforces the behavior that preceded it.

Startling a cow is aversive and can result in unintended consequences. Startle behaviors are reflexive and involuntary, resulting from novel, sudden, or intense stimuli. Although the response is quick, the adrenaline effect can last a few minutes, or the learned behavior can be repeated for the rest of her life.[5] Aversive techniques, such as yelling, intentionally striking the cow, or chasing the cow with a paddle or prod might undermine her intended escape behavior, but they make subsequent interactions even harder.

USING A ZIGZAG TO MOVE CATTLE

Those who raise beef cattle have an inherent advantage over those raising dairy cattle, because calves typically stay with their dam (Figure 7.2). Calves learn a significant amount by following the mother cow, including how to move away from humans during herding. If the mother cows were handled gently, the calves are generally easy to work.

Beef cattle raised in open spaces tend to be more reactive than dairy cattle or other beef cattle raised near humans. Intact bulls and nursing mother cows are usually more reactive than steers or dry cows, and certain genetic lines of cattle are more reactive than others. Facility design can aid in the process of humane handling,[4] but effective and efficient handling does not require it. Efficient handling depends more on technique than on equipment. Whether in well-designed facilities or in relatively open spaces, familiarizing cattle with low-energy handlers should teach them to walk during herding and processing—a much safer and more efficient way to work them.

Proficient livestock handlers realize that startling sounds and the newness of any place or space can be problematic. Any change in a pasture, pen, alley, or chute will cause the animals to investigate. Novel or new objects or noises are simply that—unfamiliar. If cattle perceive that something is new or has changed, they need time to adjust to it. A bucket can be moved or a set of pens can be rebuilt. Allowing the cattle to move through the changed chute without entrapment can make a new space no longer seem new to the cattle. Sharp and loud noises also need to be avoided. Talking is what people do, but it is merely a stimulus to a cow.

Cattle handling has been a pedestrian event since the time of early domestication, but whether humans interact with cattle on foot or mounted in some form, the ability to withdraw or slow immediately is very important. It is easiest for handlers to start learning efficient, humane cattle-handling procedures with one animal in a pasture or in a pen. They may then proceed to two animals and so on until the handlers are competent at moving the group.

FIGURE 7.2   Beef calves follow their mothers and learn herding behaviors while they are young. This avoids learning to face people, as usually happens with dairy calves. *Jeff and Keiley Banfield Red Angus Farm, Southern Pines, NC, USA. © Don Höglund (2012).*

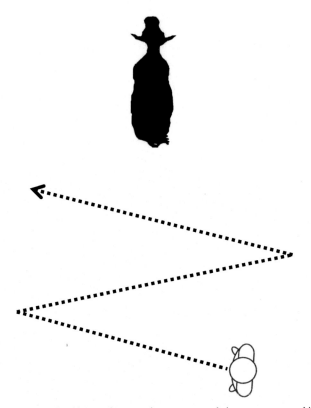

FIGURE 7.3 Handlers should use a zigzag motion to move cattle in open spaces, making sure that they go far enough to each side so that they can see the cow's eye. They work closer to calm cows and farther back from the more reactive ones. © *Don Höglund (2015)*.

The zigzag driving or herding approach is an efficient method for moving an individual cow or group of animals in open spaces. The handler is most efficient when move in each direction is continued until the handler can see the animal's eye on that side so that the cow does not have to turn to see what is going on behind her (Figure 7.3). The handler motion is usually enough to stimulate continued movement in cattle. It will be somewhat closer to the animal or herd if they are docile and farther back if they are reactive. Increased handler speed and closer encroachment are more efficient than vocal stimulus at moving docile or tame cattle.

Because cattle tend to follow the herd, handlers can take advantage of that social trait to move several individuals. In a pasture or pen, the pattern of zigzagging behind the cattle is used (Figure 7.4). It was originally described by Bud Williams as "zigzag a T to the gate." The handler moves left to right behind the cattle so that the general line of his movement is roughly perpendicular to a straight line to the goal.

The handler enters the gate, skirts the perimeter of the pen until he is opposite the gate. In larger pastures, and if the cattle are not overly reactive, walking or riding directly through the herd uninterrupted is an equally efficient route

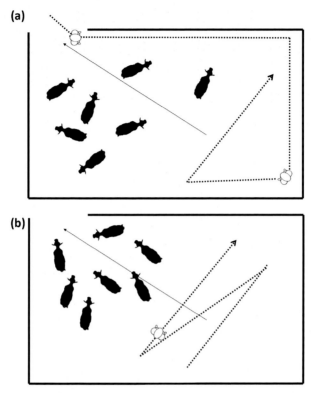

FIGURE 7.4   To move a group of cattle out of the pen, (a) the handler enters the pen and skirts along the edges to reach the side opposite the gate. (b) From this location, the handler begins a zigzag pattern in straight lines and going far enough to each side to see the outside eyes of the cattle but keeping slow, steady encroachment on the cattle to keep them moving forward. © *Don Höglund (2015).*

to the back of the herd. Then the handler visualizes a direct line to the gate and begins the zigzag motion from the outermost animal, side to side. Reactive cattle will respond to handler motion and will begin herding away from handler stimulus. The handler then moves in a zigzag pattern toward each side of the herd until he sees the outside eye of the farthest cow on that side, and then he does the same in the opposite direction. This type of negative reinforcement (moving away from a stimulus) is the cornerstone for conditioning cattle to herd, enter and exit gates, approach and navigate chutes, and accept examination by handlers. If there is more than one person working the cattle, all need to move in the same direction and with the same intensity (Figure 7.5).

Escapes happen when an animal turns away from the gate. It is often caused and then reinforced by the handler when he chases after it, yelling, or allows the animal to turn away from the gate a second or third time. It is very important to avoid adding stimuli other than just the handler's presence until it is really needed. Cattle will avoid any spot they learned to associate with pain or startle. In addition, repeated successes at escape reinforce the behavior even more.

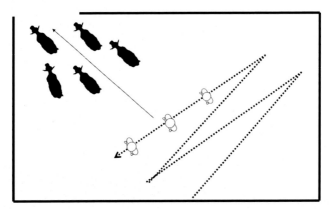

FIGURE 7.5 When more than one handler is involved in herding the cattle, all handlers need to move in the same direction at the same time and at the same speed. This provides consistent stimuli for the cattle. © *Don Höglund (2015).*

The bottom line is to allow the renegade to separate from the herd. The handler should then make an immediate change in his position farther from the renegade to give it an opening it can use to rejoin the herd. Other changes can be made to encourage it to rejoin the herd, such as making a change in the flow to the gate by adding stops in various places in the pen, changing the angle of handler's approach, and slowing the handler's encroachment to slow the herd or escapee if they are trotting. It may be necessary to use gates to sweep the herd through the problem opening or to bring portable equipment to the animal location.

For a single defector steer, a few minutes spent teaching him to herd at a walk and to park at various spots in the pen is time well spent. Begin by walking the steer around a pen, staying to the inside of the circle and far enough behind him so that the gait is a slow walk (Figure 7.6). Then direct him to a specific location in the pen. Back up slightly to stop him. Hold him in that location for a couple of minutes by using slight handler left-right and forward-back movements as needed. Then herd him to another location. Park him in various spots in the pen and finally in front of the gate. Hold him facing the opening first. Then the handler's stimulus of rocking slowly side to side or potentially back and forth will often prompt the steer to step away from the motion of the handler and through the gate (Figure 7.7). Intractable avoidance behavior may require temporary panels to be used to enclose the animal in an increasingly smaller area around the opening.

It is essential to reprompt the steer's movement through the gate just before he attempts turning away or tries other escape methods, such as jumping the panels or rushing past the handler. Efficient handlers manage cattle from a place where the animals can see them—not by adding vocal commands, arm waving, or rattling plastic or paddles. As cattle watch the handler, they will tend to arc around him, so this technique can be incorporated to help change directions at gates and alleys (Figure 7.7).

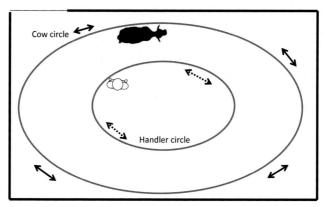

FIGURE 7.6   Reactive cattle and escapees should be retrained individually by first putting them in a pen and following them at a pace slow enough that they just keep walking to move away from the handler. © *Don Höglund (2015).*

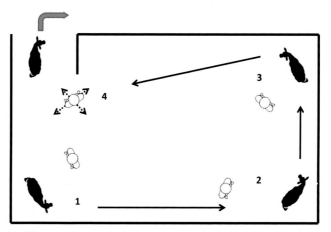

FIGURE 7.7   When reactive cattle have learned to be parked in various locations in the pen, such as positions 1–3, they are herded slowly to the area near the gate (position 4) and parked. Slight side-to-side and back-and-forth movements hold the animal in each location. The steer is then encouraged to move out the gate, arcing around the handler, as the person slowly moves back and forth in a small zigzagging forward movement. © *Don Höglund (2015).*

Handling individual or groups of reactive cattle is done the same way. First, the handler trains them to walk—always walk—and then to park facing slightly away from the handler. Hold the position. The cattle are then moved and parked at various places in the pen but always away from the gate (Figure 7.8(a)). The parking should last at least 2 min before moving to another parking spot. After 10 min, park them at the opening of the gate and keep them facing the gate. It is important to keep them facing the opening. The handler then repositions himself to take advantage of the natural tendency for cattle to arc around something

(a)

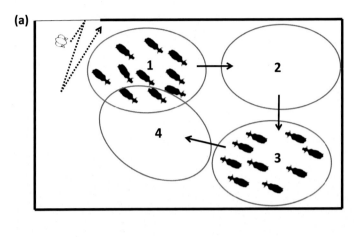

(b)

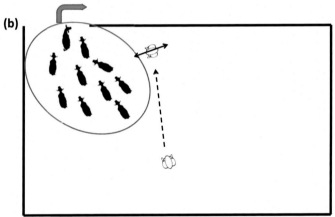

FIGURE 7.8   Groups of reactive cattle can also be taught to move in a slow walk and to park is specific locations. (a) The handler uses a zigzag herding motion to walk the group to a location where he then uses a back-and-forth and a side-to-side motion to hold them in that location for a few minutes. They are then moved to different locations within the pen or pasture. (b) The cattle are then moved to and parked near the gate, being held with the same slight handler movements. Finally, the handler moves nearer to the gate in a position that the cattle will spot as they move in an arc out of the gate. © *Don Höglund (2015)*.

they are watching (Figure 7.8(b)). If the cattle keep trying to avoid the opening, they can be herded up along a fence line that leads to the opening in a way that allows the handler to position himself so his movements encourage the cattle to continue forward as they arc around him.

SORTING PENS OR CORRALS

The most efficient position for starting movement of cattle in chutes and wider alleys is for the handler to work from the side of the animal (Figure 7.9).

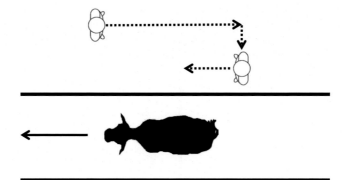

FIGURE 7.9   When working cattle in a chute or narrow alleyway, the handler needs to stay within the peripheral vision of the animal. If he starts in front of the cow and walks toward the rear, the cow will initially slow or stop. As soon as the handler passes the shoulder, animal motion starts to move forward again. Then, from a position toward the rear of the animal, the handler turns and moves forward at the same speed as the cow and parallel to her line of movement. © *Don Höglund (2015).*

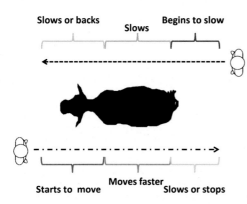

FIGURE 7.10   The speed and direction of an animal's movement is dependent on the direction of the handler's travel, speed, and position of his or her body relative to that of the cow. © *Don Höglund (2015).*

This lateral position affords the animal an opportunity to maintain continuous visual contact of the handler. Once the handler passes the shoulder, animal motion starts to move forward. From a position near the hip of the animal, the handler turns and moves forward at the same speed as the cow and parallel to her line of movement. The direction of the handler's travel, speed, position relative to the body of the animal, and distance from the cow affect her response (Figure 7.10).

Herding or driving animals in alleys or between fences and walls is done efficiently by coursing in a zigzagging fashion from behind the animal (Figure 7.11), minimizing the time spent directly behind the animal. Cattle try to see what is directly behind them and so will stop to look. By showing up on both left and right sides, the cow responds by continuing forward movement.

Cattle are gregarious, following one another until a stimulus interrupts the flow. For reactive cattle that are prone to jump, they are less likely to do so

FIGURE 7.11 Whether herding a single cow or a group down an alleyway, the handler should use a zigzag pattern going to both left and right sides so that the cattle occasionally see the person in their lateral vision. © *Don Höglund (2015)*.

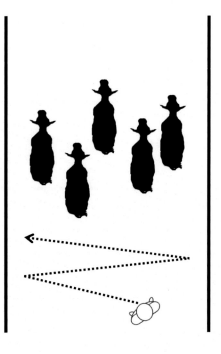

if they are prevented from seeing what is on the other side of the fence. That makes solid wall alleys, lanes, and chutes desirable. On the other hand, if the handler and the cattle cannot see each other, it is hard for the handler's presence and action to stimulate progressive animal movement. Handlers are perceived as abruptly appearing over the top of the lane, startling the cattle. The animals may even try to climb over one another in an attempt to flee the abrupt appearance. This is especially true if there is also intense or abrupt noise. It is important that if solid sides are used, a catwalk or other setup is incorporated into the design so that the cattle will not be startled.

From the dawn of domestication, cattlemen herded cattle over long and short distances without the aid of solid wall enclosures. It can be done efficiently by first teaching the animals to associate events with people, places, and experiences in the environment with calm movement. Even reactive range cattle or breeds prone to using escape behavior can be conditioned to accept handler presence and action. This is done by teaching them to walk everywhere they go through slow, quiet handling.

Efficient handlers use the "golden-rule" of cattle herding: maintain cattle motion and move in the direction of the destination. Stop-start handling methods teach cattle to move a foot or two and then stop. After as few as two or three repetitions, start-stop patterns develop. Tail jacking is a primary source of go-no-go cattle behavior. If tail jacking must be used, hold the twitch as the animal moves all the way through the chute to the end. In some cases, cattle move a

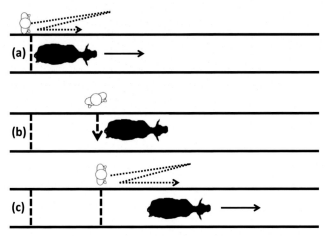

FIGURE 7.12 Moving resistant cattle in a chute or alleyway without solid sides involves (a) first putting a see-through blocking gate behind the steer as he enters the chute. Then the handler initiates movement by moving forward at a slight angle away from the steer's head. He or she then turns and walks toward the rump to initiate forward movement. (b) At the rump, another blocking gate is inserted (c) before the series is repeated. © *Don Höglund (2015).*

foot or two and stop, regardless of the efficiency of the handler. Rump gating in a chute or alleyway often helps to maintain the ground gained by each step forward. Panels, rails, and gates should all be see-through equipment so that handlers and cattle can see each other. To move an individual cow through an open-sided chute, the handler first walks at a slight angle going toward the chute sides, moving from the cow's head toward her rump (Figure 7.12). This causes the cow to move forward to increase her distance from the handler. The handler then reverses direction, staying next to the rump. When the cow stops (or periodically, in case she will), a block gate is put behind her so she cannot back up. Then the handler angles away from the animal until he reaches a position at about her neck. The animal cannot move back, and the handler again moves parallel but opposite the cow's line of motion, slightly angling toward her hip (Figure 7.12). Each progress is locked forward by the blocking gate so that the animal cannot back up. Excessive stimuli should be avoided and can actually be counterproductive.

THE REDIRECTION PEN

Cattle have a tendency to try to return to familiar spaces. That tendency, coupled with the behavior to move away from the handler, can be used to get cattle into chutes or alleyways with a redirection pen (Figure 7.13). This design was popularized by Bud Williams, resulting in its other name—the "Bud Box." Any business that processes cattle can benefit from having a redirection pen, particularly beef breeding operations and processing facilities where frequent weighing and transport handling are necessary.

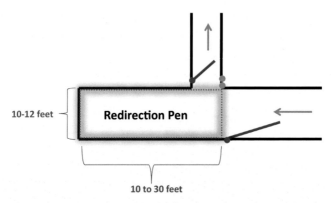

10-12 feet

Redirection Pen

10 to 30 feet

FIGURE 7.13   The design of a redirection pen (also known as a "Bud Box") makes it easier to direct cattle toward unfamiliar areas. The pen is 10–12 feet wide, 10–30 feet long, and has an exit chute that is the width of a single animal. Note that the location of the gate hinges (●) and latches (●) and the direction the gates are open are very important. © *Don Höglund (2015).*

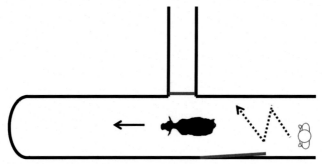

FIGURE 7.14   Cattle are walked into the redirection pen using the zigzag pattern. The rounded end of this pen will aid in turning the cattle back toward the gates. © *Don Höglund (2015).*

Pen design is important, especially where the latches and hinges are located. The latch of the entrance gate needs to be next to the opening of the intended exit lane. Cattle return to where they enter the pen and when they do, the exit lane opening needs to be immediately available.

The dimensions of the pen should be 20–30 feet long and 10–14 feet wide for beef animals. While the far end of the pen can be flat, an arcing barrier is valuable for directing the cattle back to the gate. The fencing should be the pipe-rail or livestock panel type with round tubing used throughout. Solid wall fencing has the propensity to startle cattle as handlers seem to pop up suddenly. A "catwalk" is useful if the walls are currently solid.

The exit lane gate to the transport, alley, or working chute should initially be closed. Handlers walk the cow or group into the redirection pen using the zigzag from side to side at the rear of the animal method (Figure 7.14). The handler securely closes the main entrance gate, opens the exit lane gate, and stands quietly inside the pen on the hinged side of the exit lane gate (Figure 7.15).

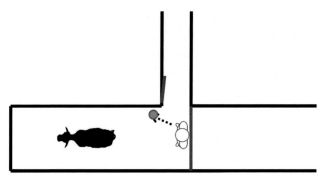

FIGURE 7.15   Once the animal is in the redirection pen, the handler closes the entrance gate, opens the exit gate, and moves to an area near the exit gate hinges. © *Don Höglund (2015).*

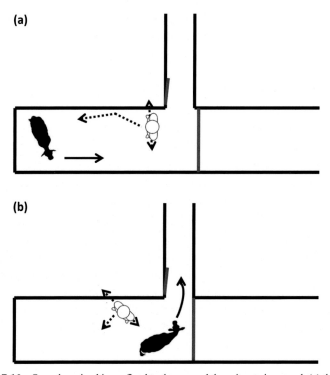

FIGURE 7.16   Once the animal is confined to the pen and the exit gate is opened, (a) the handler moves slowly down the side of the pen and uses a side-to-side rocking motion to stimulate the animal to move back toward the opening. (b) Because of the handler's position, the animal will arc around him and enter the exit alley. © *Don Höglund (2015).*

The animal will start to arc around the handler as it moves back toward the main gate (Figure 7.16). When it reaches the closed main gate, it spots the open exit lane and is on its way to the weigh station, transport, or working chute. The only movement by the handler may be a short move along the side of the pen toward

(a)

(b)

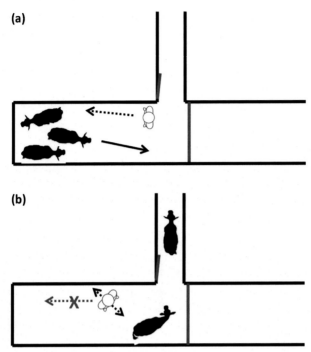

FIGURE 7.17   (a) The same technique used on a single animal in the redirection pen is used for groups of cattle. Once the cattle pass the handler, (b) he or she does not continue to the far end of the redirection pen, but instead faces the last animals and uses slight motions to continue their momentum. © *Don Höglund (2015)*.

the back end and a slight side-to-side rocking motion. The change of sound made by the hooves or the physical appearance along a pathway when the steer enters a transport can cause it to balk. This problem should be anticipated and prevented by altering the surface of the transition from pen to exit lane.

Unfamiliar places can be made familiar by having animals trace the route earlier without incident, trapping, startle, or pain. Simply walking cattle through alleys, chutes, or a redirection pen setup removes the unfamiliarity. Then when it is time to use the "box," there is less chance of problems.

Groups of cattle are managed the same way in redirection pens as are individuals. Efficient handler position keeps handlers at the side of the animals, exhibiting slow movements, and allowing cattle to look where they are going (Figure 7.17). Handlers that move behind the cattle and those that stand by the hinges of the entrance gate cause cattle to look back, and this slows momentum of the animals entering the exit lane.

The "sorting pie" is a modification of the redirection pen with multiple exit gates leading away from a round "box" (Figure 7.18). This configuration is particularly useful for sorting cattle. Management of single cows and multiple cattle can be affected by handlers working from the center of the pen toward

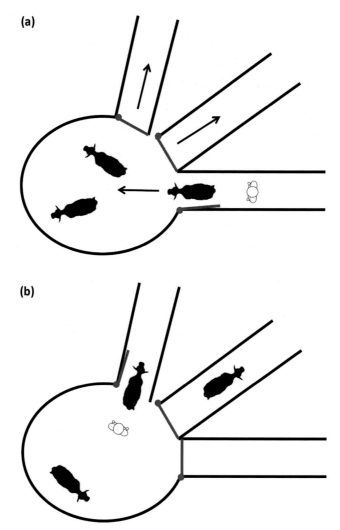

FIGURE 7.18 A "sorting pie" is a round version of the redirection pen, in which cattle can be directed into a number of different exit points based on where they need to go, such as a transport, chute, scales, and holding pen. Note the gap in the sides of the sorting pie is a pedestrian walk-through area. © *Don Höglund (2015).*

the exit-lane hinge. Handlers should check to be sure that the exit-lane gate is closed after the animal has exited the round redirection pen.

Of particular importance to efficient, humane handling of beef cattle, whether they are entering or exiting a redirection pen or being worked elsewhere, is the avoidance of all unnecessary stimuli. This includes untrained dogs, cats, tractors, and humans. Electric prod devices and poker rods are often abused and are best never used.

THE USE OF A TRAP

Another method for working cattle for any purpose is the use of a corner trap or a chute trap (Figure 7.19). In a corner trap, the animal is blocked into the corner of a pen using a panel or similar barrier. For the chute trap, animals are driven forward to a head catch and then confined with a restraint placed behind them. Once the animal is trapped, the handler begins to approach the animal very slowly and quietly at zigzagging 45° angles, trying not to step ahead of the shoulder or behind the hip. Reactive animals typically attempt escape or avoidance behaviors. It is critical that while this animal reaction is underway, the handler needs to continue slow, purposeful encroachment. When by chance the animals stops moving, the handler should immediately stop moving forward and even take a small step back. This teaches the animal to associate its action with a specific human reaction that relieves stress—negative reinforcement. Repeating the process allows the handler to get closer each time, with the eventual goal of touching the animal. If the animal is moving while touched, the handler should not remove the hand until the animal has stopped fidgeting. The basic concept is that if the animal is fidgeting, the approach or touch is continued. When the animal stops fidgeting, the behavior is reinforced by the person stopping the touch or approach, and even stepping back a very short distance. It is important not to reward the unwanted behavior.

FIGURE 7.19   The use of a trap is another way to teach cattle to accept close proximity to humans. (a) A panel can be used in a pen or (b) a head catch and back stop used in a chute. The handler approaches in a zigzag motion, staying between the animal's shoulder and hip. Forward motion continues as long as the animal fidgets, but stops as soon as the animal holds still. © *Don Höglund (2015)*.

REFERENCES

1. Chabris CF, Simons DJ. *The invisible gorilla; how our intuitions deceive us.* New York: Broadway Paperbacks; 2009. pp. 306.
2. Dobie JF. *The longhorns.* Austin: University of Texas Press; 1940. pp. 440.
3. Gilger BC. *Equine ophthalmology.* New York: Elsevier; 2010. pp. 536.
4. Grandin T. *Humane livestock handling: understanding livestock behavior and building facilities for healthier animals.* North Adams (MA): Storey Publishing; 2008. pp. 227.
5. Lansade L, Bouissou MF, Boivin X. Temperament in preweanling horses: development of reactions to humans and novelty, and startle responses. *Dev Psychobiol* July 2007;**49**(5):501–13.
6. LeDoux JE. Coming to terms with fear. *Proc Natl Acad Sci* February 25, 2014;**111**(8):2871–8.
7. Le Neindre P, Guémené D, Arnould C, Leterrier C, Faure JM, Prunier A, et al. Space environmental design and behavior: effect of space and environment on animal welfare. In: *Proceedings of the global conference on animal welfare: an OIE initiative.* February 23–25, 2004. p. 135–48.
8. Mader DR, Price EO. The effects of sexual stimulation on the sexual performance of Herford bulls. *J Anim Sci* August 1984;**59**(2):294–300.
9. McGreevy P. *Equine behavior: a guide for veterinarians and equine scientists.* New York: Saunders; 2004. pp. 369.
10. McLean AN. The positive aspects of correct negative reinforcement. *Anthrozoös* 2005;**18**(3): 245–54.
11. Pajor EA, Rushen J, de Passile AMB. Aversion learning techniques to evaluate dairy cattle handling techniques. *Appl Anim Behav Sci* September 2000;**69**(2):89–102.
12. Phythian CJ, Cripps PJ, Michalopoulou E, Jones PH, Grove-White D, Clarkson MJ, et al. Reliability of indicators of sheep welfare assessed by a group observation method. *Vet J* January 20, 2012;**193**(1):257–63.
13. Tinbergen N. *The study of instinct.* Oxford: Clarendon Press; 1955. pp. 254.
14. Veissier I. Observational learning in cattle. *Appl Anim Behav Sci* January 1993;**35**(3):235–43.

Chapter 8

Swine Handling: Practical Applications of Science

The most damaging phrase in the language is: "It's always been done that way."

Grace Hopper

John was a city man most of his life, but when he was 52, he inherited about 50 acres from an uncle. He moved his family there and even built a house on the property. Then he got the urge to be a "farmer," so he bought a tractor and all the "utensils," as he called the implements. John was not afraid to tackle anything, even if he was not exactly sure how to do it. There were some cows on the place already, so John thought pigs would make a nice addition. He made a small pen and then bought five quarter-grown pigs from a neighbor. When he got tired of hauling water from the house to the pen several times a day, he decided to lay a waterline to make his life easier. John spent an entire weekend digging the trench by hand and then laying a plastic pipe to the pen. When he returned home Monday afternoon from his "real job," John was shocked to see water everywhere and two of the pigs running around with short sections of plastic pipe in their mouths. It seems they had a way of finding their own entertainment.

Handlers and farm animals typically interact several times every week. Each interaction between humans and swine creates an opportunity to condition the animals for productive outcomes and to accept human presence and action during day-to-day handling. Humane pig handling involves understanding how pigs learn and how to manipulate that relationship for optimum current and future outcomes.

REACTIVITY AND STRESS ARE RELATED

Handling includes being touched by humans and herded to some destination. Initial touch-handling is often associated with negative interactions, such as being lifted, examined, castrated, tooth clipping, or injected. While these things have to be done, gentle handling is less stressful. It has been shown that increased positive human contact will lower brain beta endorphin levels and heart rate, which are both indicators of stress.[1] This is a direct indication that skillful handling can lower stress and

Efficient Livestock Handling. http://dx.doi.org/10.1016/B978-0-12-418670-5.00008-1

179

improve the quality of life for the animals. It is no surprise that calm, consistent animal handlers often have minimally reactive animals.

On the other hand, squealing in pigs is a sign of reactive animals responding to a strong stimulus. If the pig is released from the handling while squealing, it can associate squealing with successful survival behavior—a reward. Releasing the pig only when it has quieted will condition it to associate quiet with imminent release.

In the modern swine production units, piglets do not have an opportunity to learn herding behavior by following their dam. Once gathered and placed in group pens, weaned pigs are often subjected to handlers attempting to capture them or herd them from place to place. Teaching swine to herd calmly and progressively toward the intended destination is easily taught and would be valuable on most swine farms.

Humans cannot know what a pig is thinking, feeling, or wanting. Rather, concentrating on what the pigs are actually doing during handling allows a handler to make adjustments to accomplish what is needed. For instance, pigs that are running from handlers, avoiding places and objects in their path, or facing the human are not herding optimally. Identifying the stimulus that causes the unacceptable behavior means the problem can be stopped.

Teaching animals to herd is not complicated. Instead of focusing on paddles, noisemakers, herding boards, and solid-wall encasements, handlers are better off by spending time teaching the animals to herd from place to place in their weaning pen, along alleyways, and elsewhere on the farm. Learn the hallmarks of when to reduce the human presence as a stimulus on the animal in order to efficiently condition them to herd at a walk instead of a trot.

In recent years, consumers have started to take an interest in how food animals are raised because of the horror stories that show up on YouTube videos. While many of those are staged, the public has no way to determine it. At the same time, most swine producers and researchers who use pigs really care about doing what is right for the animals in their care. There is a real interest in learning the newer techniques that improve humane handling. Because animal and human contact is often cited as the leading cause of injuries on the farm, the safety of workers and animals is a prerequisite of any humane handling education and training.[2,3]

Though livestock and human relationships are similar the world over, it is essential to recognize that every livestock operation is unique. That uniqueness is amplified by the varying adaptations of modern livestock production. Even in the event of identical facility designs, the behavior of the handlers is surely different, so pigs learn different things. This means that careful observations of what the pigs are doing and anticipating how they might react to outside stimuli is important in determining what step is next in handling.

Understanding efficient, humane swine handling techniques requires practical knowledge of pig behavior, the senses, and the types of learning. Strong, positive contributions to the appropriate management of swine behavior demands precise and unambiguous language and technique. Anthropomorphisms and trending concepts regarding swine behavior, temperament, and training have ignored well-established principles of learning. The practical power of low-energy stock handling must rely on education and observation instead of the proliferation of misinformation and pseudoscience.[5] As an example, "psychological pressure" implies humans can know the subjective consciousness in swine, which of course we really cannot.

LEARNING IN SWINE

As is true for most livestock species, pigs learn primarily by trial and error and the associated reinforcement. A *stimulus* is any detectable change in an animal's environment and any *response* is a resulting behavioral or physiological reaction.[4] Learning in livestock happens when a consistently applied stimulus results in the probability of a predictable behavior. The concepts of learning, discussed in Chapter 4, are applicable to swine and are briefly reviewed here.

Classical conditioning is where the pig has no control over the events but responds to a new stimulus by associating it with an old stimulus. For instance, the pig learns that the bucket brings grain, and soon she learns to associate the sound of grain being poured into the bucket with being fed. Operant conditioning is about the consequences. This means the pig has some control of the outcome because she finds it to be desirable or not.

Timing of the stimulus, consistency of the technique, and immediate reinforcement are critical elements in learning. Pigs respond best to instantaneous rewards (positive reinforcement) or instantaneous relief from a stimulus (negative reinforcement). It is important here to remember that negative does not imply bad, but rather that there is relief when a noxious stimulus was lessened. For example, if a person steps toward a pig and it moves away, the encroachment stimulus was reduced when the pig moves—a reward that reinforces the behavior of moving away from the person. Negative reinforcement underpins most swine handling, though it is usually subtle.

Farm facility flow and human understanding of learning in pigs will assist handlers in reducing adrenaline, endorphins, and heart rate levels in all farm animals. Less stress for the animals also means less stress for the handlers.

On their own, pigs quickly learn to manipulate food and water devices through trial and error. This interest in exploring their environment is also used to test preferences. Pigs turn barn fans and lights on and off. If the outcome is rewarding, pigs can learn to repeat it. Researchers then increase the difficulty of obtaining the reward to see how hard the animal will work for the outcome. For example, a gate may be weighted, making it harder to open to get to food, or a light switch made more difficult to move. The amount of effort the pig is willing to use to get to the food or turn on the light indicates how important it is to the animal. However, just because something is important to the animal does not necessarily mean it is good for the animal. A person might work significantly harder to get a piece of chocolate than a piece of broccoli. Preference tests do give a lot of information, but the results must also be carefully interpreted.

FIGURE 8.1 Pigs look at a novel stimulus like the shadow in the opening and stop moving. This disrupts smooth forward movement, so it is important to avoid novelty in the environment and during handling.

Pigs are curious by nature and will stop to look at new things. This will interrupt a smooth flow of moving a group forward (Figure 8.1). If the stimulus is sudden, it often causes the pigs to startle. Once this has occurred, the pigs will be under the influence of adrenaline for at least 20 min. During this time, the animals are more difficult to handle. To prevent human-associated startle responses, slowing overall handler action and eliminating quick movements is necessary. Moving swine is done most efficiently if they can actually visualize the handler at all times. A rocking motion devoid of quickness, with or without herding boards or paddles, is the best technique for a handler to use to move swine. Electric devices, loud noises, and rapid movements are all associated with adrenaline release and should only be used in emergency situations. In all handling situations, it is most efficient to increase or decrease stimuli as one would climb a ladder or operate a rheostat: one increment at a time. Handlers who employ methods that are all on or all off, or even skipping rungs on the energy ladder, create startle in pigs. Seamless, continuous, rheostat-like control of handling stimuli is best when micromanaging swine behavior.

HERDING PIGS

Pigs have a relatively large blind area behind them, so handlers must be careful to not spend a lot of time there, causing the pigs to turn to see the handler's location and breaking forward momentum. In addition, the anatomy of a pig's neck makes it very difficult to simply look back, and the forward path becomes an arc instead (Figure 8.2). Efficient handling keeps pigs moving in the intended direction. If they are allowed to arc or turn around, they learn to arc or turn around. That action interrupts smooth herd flow. Knowing this, efficient handlers attempt to herd swine from positions that allow the pig nearly continuous visual contact with the handler. This means that because of their visual limitations, the best position to herd pigs is from the side, between the point of the shoulder and the middle of the back (Figure 8.3). Positions ahead of the shoulder will result in the pig turning back. Farther behind will cause the arc. As it has been described for the herding of other livestock, a quiet, zigzagging pattern is used both from the side and, when necessary, from behind (Figure 8.4). It is the predictable motion of the handler and continuous, or nearly continuous, visual contact that provide the stimulus for pig locomotion.

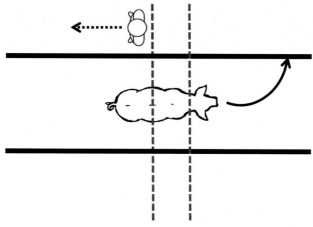

FIGURE 8.2 When a handler attempts to move a pig forward in an alleyway, they will walk toward its rear, but if they go farther back than midway, the pig will arc as it tries to keep visual contact with the person. This is inefficient. © *Don Höglund (2015).*

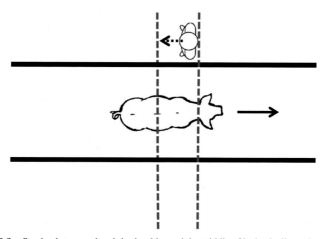

FIGURE 8.3 Staying between the pig's shoulder and the middle of its back allows the pig to keep visual contact with the handler as it responds to the closeness by moving forward. © *Don Höglund (2015).*

Past animal experiences, current handler stimulus, animal reactivity, age, current location, and destination are factors determining the level of handler action required to move one or more animals. When herding swine in pastures or large paddocks, a straight-line, zigzag pattern is most efficient (Figure 8.5). This is the same as for other livestock. Except for brief periods when the handler crosses the midline behind the animal(s), predictable, straight-line encroachment at roughly 45° angles prompts swine to move forward. The handler must be sure to move far enough to each side so that he can easily be seen by the pigs.

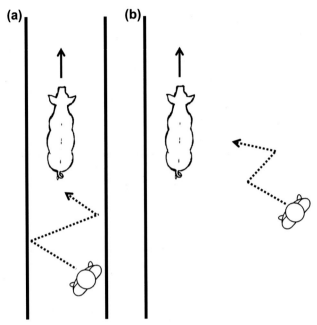

FIGURE 8.4   A zigzag pattern, where the pig can see the person on each side of its body, is useful for moving the animal straight forward when it is (a) in a chute or (b) along a fence or wall. © *Don Höglund (2015).*

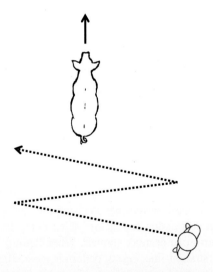

FIGURE 8.5   Herding a pig in an open area requires that the handler move in a zigzag pattern that is wide enough that the pig(s) can see the person on either side. The handler should move in straight lines that connect at approximately 45° angles. © *Don Höglund (2015).*

Consistency in handler-to-handler technique, rate, timing, and angle of approach conditions the animals about what to expect next. Changing handler action, inconsistent methods between handlers, and unneeded stimuli will cause the pigs to either face the stimulus or collide with each other.

When moving swine toward a gate or corner, handlers will find that the zig-zag pattern is again useful (Figure 8.6). The lines of handler movement should be straight, wide, and approximately perpendicular to a direct line to the goal. Arching movements, instead of straight lines, make it hard for the pigs to focus on the handler and should be avoided.

Once the handler has the pigs at the gate, if the pigs resist, the handler can move along the herd to the lateral position on the side of the animals toward where the pigs are intended to go (Figure 8.7). Working the pigs from this position causes them to move in a concave arc as they keep an eye on the handler.

THE REDIRECTION PEN

If swine are reactive, an adaptation of the cattle redirection pen can be particularly useful. With it, handlers can teach the pigs to move calmly or load into trucks or trailers. These are also useful to help teach handlers where to stand, how to move the animals, and most importantly, how to reduce their energy around reactive swine. The redirection pen, also called the "Bud Box" after Bud Williams, who made it popular, is helpful for loading and sorting livestock. Facility design may have solid walls or a fence/pipe sides. The important thing is that the pigs need to be able to see the handler, rather than be surprised by a pop-up figure. The location of hinges and latches is important (Figure 8.8). Owners and managers should use facility innovations that work most efficiently for them.

Once in the redirection pen, pigs tend to return to the spot that they entered, but they find the exit gate open instead. To work with the animals, the handler must shut the entrance gate and open the exit one, which leads to the new destination. Since the design and operation of the redirection pen is similar to all species of livestock, illustrations of its use are shown in Chapter 6 (Figures 6.25–6.30). After opening the exit gate, the handler stands by the hinges of the exit gate and then slowly moves along that pen wall to start the pigs moving back toward the entrance gate. Because the exit gate is the only one open and the handler is kept in view, each pig will arc toward the exit and move out.

For large facilities where sorting pigs is common, a sorting pie can be used. This is a circular penning variation of the redirection pen with multiple outlets (Figure 8.9). Handlers should attempt to move swine from a position inside the curve or arc of the pig's movement. This allows the animal ability to keep at least one eye on the handler at all times. Circular arrangements are used most efficiently if the pigs are familiarized with following a circular fence when younger.

(a)

(b)

(c)

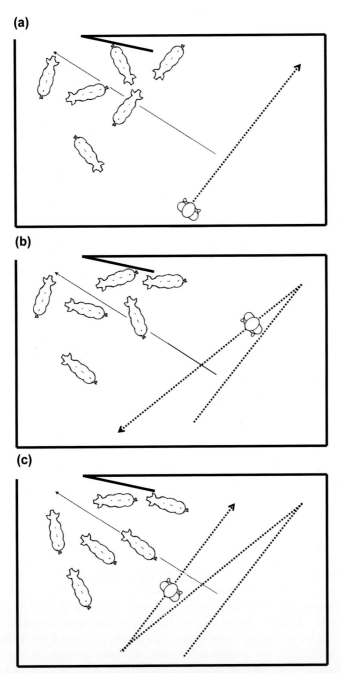

FIGURE 8.6   To herd pigs to a specific location, (a) the handler begins a zigzag pattern that is perpendicular to a direct line to the goal. (b and c) The lines of handler movement should be straight and meet at approximately 45° angles. © *Don Höglund (2015).*

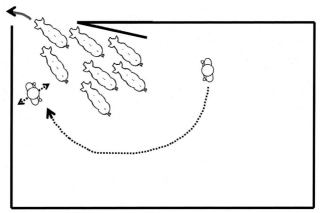

FIGURE 8.7 To direct swine movements as they walk out of a pen, the handler will move to a position on the side where the pigs are to go. Pigs will arc in the direction of the handler as they try to keep him in their lateral vision. © *Don Höglund (2015).*

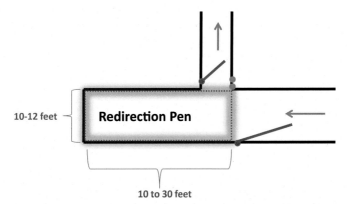

FIGURE 8.8 The design of a redirection pen (also known as a "Bud Box") makes it easier to direct pigs toward unfamiliar areas. The pen is 10–12 feet wide, 10–30 feet long, and has an exit chute that is the width of a single animal. Note that the location of the gate hinges (●) and latches (●) and the direction the gates open are very important. © *Don Höglund (2015).*

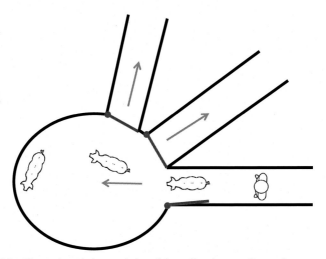

FIGURE 8.9 The sorting pie is a variation of the redirection pen. It contains an entrance and multiple exit points leading to loading ramps, chutes, scales, or other pens. © *Don Höglund (2015).*

TEACHING PIGS TO HERD

Large commercial swine breeding operations are usually built with alleyways that allow animals to be moved to a variety of locations as needed. When that type of arrangement is not available, being able to herd pigs efficiently can still occur. The key is to teach young pigs correct responses before they have the opportunity to learn undesired ones. Conditioning takes remarkably little time in the work day, and it proves tremendously valuable in teaching swine to accept human presence and action.

Begin conditioning young swine to herd by having the handler enter the weaning pen and stimulating the animals to move toward the feed (Figure 8.10). Expect that pigs will run and exhibit mock fighting behaviors for the first few minutes of handling, but the handler should slowly move left to right behind them to encourage the directional movement. Keep them near the feeders for at least 2 min. In this manner, handlers can reapply the prompt to stay near the feed bunk through the first few pig avoidance behaviors.

The goal in these initial sessions is to teach the pigs to face away from handlers. This is done using negative reinforcement by taking a step or two forward and then stepping back the moment the pig turns away.

After 2 min of holding the pigs at the feeders, the handler should move them to another spot in the pen (Figure 8.11). The handler should come down the side of the pen opposite of the intended direction of movement and use a side-to-side rocking motion or slight zigzag pattern to initiate the herding. Hold the pigs in the new position for a few minutes. Repeat this moving and parking process until the pigs walk throughout movement. Then, leave the pigs at the feeders and exit the pen. Repeat this maneuver for three days and then repeat it in 30 days. Handlers find that pigs have a remarkable memory for herding and parking.

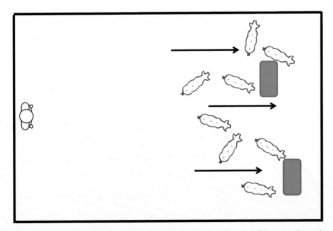

FIGURE 8.10   To begin conditioning pigs to herding, the handler should move the animals toward the feeders and hold them in that area. This is the first lesson in teaching the pigs to face away from the handler. © *Don Höglund (2015)*.

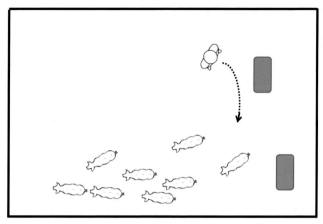

FIGURE 8.11   After a few minutes near the feeders, the handler should use a side-to-side motion to move them to another area of the pen. This is repeated several times to get the animals used to moving slowly in response to handler cues. © *Don Höglund (2015).*

Incline Training

When young pigs have learned to herd, it can be very useful to get them used to objects and locations that they will encounter later. As an example, loading ramps are commonly used when moving pigs, so a mock ramp can familiarize the animals before a ramp is actually needed. This ramp can be placed in a weaning pen. It should lead to a flat elevation that is up to 2 feet off the floor, and then it goes down again. It needs solid walls that are about 2 feet high to keep piglets from jumping or falling off. A small box can be placed on the top to introduce unfamiliar objects for the piglets to investigate. Food in the box will reward the piglets for using the ramp.

Once or twice a day handlers should herd piglets up and over the ramp. This accomplishes two objectives: it teaches the pigs to herd by human action, and it gets them used to changes in elevation. A slow, quiet, consistent handler stimulus will allow the pigs to learn how to manage funneling and ramps (Figure 8.12).

SWINE HANDLING

Swine in laboratory or biomedical confinement typically live longer than those in agriculture settings. These animals are invaluable to biomedical research, so it pays to condition them to accept human presence and actions (Figure 8.13). Lessons in handling are, however, not restricted to the laboratory setting. They are useful for any pig. It takes time and effort to humanely train pigs to herd, be alone, accept destabilizing lifting, and remain quiet and motionless. They learn by the association of events and outcomes.

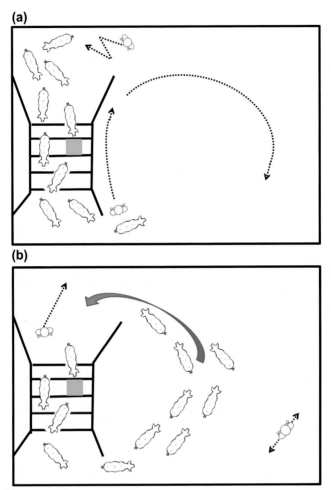

FIGURE 8.12 Placing a mock ramp and elevated platform in a pen allows handlers to teach the pigs to herd onto and over objects. (a) A box on the elevated portion is a diversion and can also be used to place novel objects. (b) Repetitions of going over the ramp make loading the pigs easier when they are to be moved. © *Don Höglund (2015).*

FIGURE 8.13 Pigs raised as labora-tory animals also benefit from learning how to be herded. *Photo compliments of Bart Carter, Southwestern Medical Center; used with permission.*

Examination of the behavioral signs of swine helps handlers discover the events in the environment that are likely to trigger reactive behaviors. To the greatest extent possible, eliminate stimuli that are not absolutely necessary. When eliminating a stimulus is not possible, teach the animal to become nonreactive when it encounters the stimulus. For example, handlers should always wear the same attire. On farms, the coveralls should be the same color. In laboratory facilities, handlers should wear their typical laboratory clothing, including gloves.

At some point, pigs that will need to be physically handled must be trained to accept it. It pays to condition piglets to the specific ways they will be handled. Initially, in the pen, handlers should keep their hands to themselves for several days, even when piglets investigate them. Sitting on a stool or squatting reduces the handler's size and helps reduce the time before the pigs investigate. Then, the handler can start getting the piglet used to being touched. Slow movement toward touch and slow withdrawal are important, if only because they are predictable and represent less stimulus. Touching starts with the back of the hand first to create a single point-of-contact stimulus rather than the four points of the fingers. After the pig readily accepts the back of the hand, fingers can be used to condition piglets to accept figure pressure. The goal is to create a stimulus and motion without precipitating shrill vocalizations, so everyone working with the animals needs to be educated about proper procedures. Pushing piglets from side-to-side gently conditions them to accept increasing touch pressure.

Lifting piglets should come only after the pig accepts touch without shrill vocalizations. Being lifted and restrained takes away the animal's ability to flee. It is also often associated with pain. It is no wonder that pigs attempt escape behavior and loud squealing. Lifting inches off the ground is eventually replaced by greater heights. Each step is done slowly and gently to be sure the piglet remains calm and nothing aversive happens. Only then should the next step come. The handler can also feed small amounts during each handling lesson to give a positive association to the handling.

Handlers need to be prepared for any behavior, and they should try not to repeat a stimulus that promotes an undesired behavior. If those unwelcome behaviors work for the animal, they are likely to be tried again. It is important to understand that releasing a squealing or struggling pig conditions the pig to squeal and struggle. From the first handling onward, only release the pig when it has quieted and calmed. Do not just drop it. Release the piglet and step away from it rather than having the piglet run away from the handler. If a handler enters a swine-holding facility and a significant number of the animals move away from the entrance, there is a handling problem at that facility. If a stimulus promotes an unwelcome behavior, change the stimulus. Do not repeat it.

In a broader sense, it is important to be aware of what is physically happening to the pig when it is restrained and then poked, prodded, or stuck with sharp objects. Each experience will likely be remembered, and restraint will come to be associated with pain, places, and human handlers. Keeping this in mind helps handlers to slow down just before touching and restraining the animal.

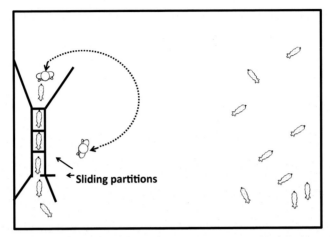

FIGURE 8.14   A catch alley can be used to treat individual piglets without having to physically catch each one. Once the piglets enter the alley, a sliding partition is placed behind them to prevent them from backing up and to stop forward motion of the next one in line. © *Don Höglund (2015).*

When pigs are involved in medical procedures or testing, giving them something positive immediately after they settle down is encouraged to teach the pig that bad things can be associated with good outcomes. Even small amounts of food, especially if it is sweet, are high value rewards in most cases. Food can also be used to hide medication.[6]

Catching Individual Pigs

Pigs can be trained to enter an alleyway that is similar to those used to restrain cattle so that when restraint becomes necessary, it is not stressful. This eliminates the need to capture piglets using the hind-leg grab or snout-snare methods. Once all materials, supplies, and medical equipment have been readied, herd the animals into the catch alley and isolate each animal by virtue of the individual sliding gates or doors (Figure 8.14). To work on a piglet, approach slowly. Touch it with the back of the hand, turning gently to the fingers. Slip the arms under the piglet's trunk and lift the animal with full support of its chest and the belly. If possible, the piglet should straddle the handler's forearms. When the procedure is done, the piglet is released only when it has quieted and stopped squirming.

Never grab swine by the ears or the front legs. These appendages can be damaged by the strain of escape behavior.

Older swine can be restrained in a catch alley or against a fence or solid wall for quick procedures like injections. For longer procedures, slings are handy to use when a pig must be restrained. It is also useful to gently rub the pig behind its ears and talk softly to it. Both can be effective calming techniques.[6]

REFERENCES

1. Geers R, Janssens G, Villé H, Bleus E, Gerard H, Jamssens S, et al. Effect of human contact on heart rate of pigs. *Anim Welfare* November 1995;**4**(4):351–9.
2. Langley RL, Morrow WE. Livestock handling—minimizing worker injuries. *J Agromed* July 2010;**15**(3):226–35.
3. Lindahl C, Lundqvist P, Norberg AL. Swedish dairy farmers' perceptions of animal-related injuries. *J Agromed* 2012;**17**(4):364–76.
4. McGreevy P. *Equine behavior: a guide for veterinarians and equine scientists*. New York: Saunders; 2004. pp. 412.
5. Shermer M. *Why people believe weird things: pseudoscience, superstition, and other confusions of our time*. New York: Freeman & Co.; 1997. pp. 306.
6. Swindle MM. *Technical bulletin: handling, husbandry and injection techniques in swine*. http://www.sinclairresearch.com/Downloads/TechnicalBulletins/Handling,%20Husbandry%20&%20Injection%20Techniques%20in%20Swine.pdf, downloaded November 19, 2014.

Chapter 9

Sheep and Goat Handling: Practical Applications of Science

It is the duty of a good shepherd to shear his sheep, not to skin them.

Tiberius

A well-known, but not well liked, animal science professor was going to use sheep in his upcoming research project. He had just spent two hours in the July sun trying to load them in a stock trailer to haul them to his research facility, but the only thing he had accomplished in that time was getting the animals really hot and tired as they continually circled the dusty pen. He was tired too. His voice was giving out from all his yelling and cursing. His hat was crumpled from hitting the ground and then being kicked. About this time, a graduate student came by who had grown up with sheep and actually took care of the university's sheep center. The student asked Dr Know It All if he could help, to which came the reply, "Those stupid sheep! I can't get them loaded." The young man stepped into the pen and gently picked up one ewe, carried her into the trailer, and tied her in the front. Then he quietly walked through the flock, slowly circling them toward the open trailer door. As the last ewe stepped in, the student closed the trailer door and then went to the front and untied the first ewe. Now both relieved and angry that a "mere" graduate student had been successful when he had not, Dr Know It All said, "That's what I hate about sheep. They are so stupid." The graduate student quietly walked away saying, "That's what I like about sheep."

While sheep and goats are small ruminants and are often spoken about in the same breath, they are genetically and behaviorally different. That means that techniques used for handling animals of each species will need to take into account those differences. Since humans can never assess what or how an animal thinks, we must rely on assessing the behaviors and the environment to develop handling methods that are safe, efficient, productive, and humane.

COMPARATIVE BEHAVIORS

Behavior in animals is defined as anything an animal does that can be observed.[1] It is a product of the species' biology coupled with learning from

Efficient Livestock Handling. http://dx.doi.org/10.1016/B978-0-12-418670-5.00009-3

195

TABLE 9.1 Primary Differences between Domesticated Sheep and Goats

Characteristic	Sheep	Goats
Biologic differences	54 chromosomes	60 chromosomes
	Wool (some sheep strongly resemble goats)	Hair (some goats strongly resemble sheep)
	Horns are thicker and curved	Horns are thin, relatively straight, and upright
	Tail hangs down	Tail is held up
	Mane	Beard ± wattles
Eating behavior	Grazers	Browsers
Fighting behavior	Rams head butt	Bucks rear up and come down with their heads
Dominance ranking	Present but not obvious	Very obvious
Play behavior	Lambs tend to stay with the flock	Kids jump and climb on elevated objects

past experiences. Using observations, some of the differences between the two species become obvious (Table 9.1).

Two behaviors are the most obvious in their differences: eating behavior and social behavior. Relative to eating behaviors, both are herbivores, but sheep are grazers and goats are browsers (Figure 9.1). As foragers, sheep show preferences for grasses, clovers, and especially broadleaf plants. Because they can tear grasses off very close to the ground, frequent pasture rotations are necessary. It was this trait that resulted in the bad blood between cattle raisers and sheep men in the Old West. Sheep could eat where cattle had grazed because cattle could not tear off grasses as close to the ground. However, the cattle could not follow the sheep.

As browsers, goats prefer the tips of woody shrubs and the tops of plants, but their diet is actually quiet varied among the plant varieties. They are often used to clean up areas overgrown with vines and weeds. In addition, their curiosity allows them to investigate and mouth other materials, but unlike their reputation, they do not eat tin cans or cardboard. This browsing eating pattern has also resulted in goats that like to climb. Whether standing with their front feet on a fence, standing on a stump, or actually in a tree or shrub, goats are more likely to use their curiosity to investigate possible feeding sites. It is also expressed as the "king of the mountain" play behaviors of kids.

The second major behavioral difference between sheep and goats is in their social behaviors. Sheep are very tightly bonded to the flock and will show a number of distress signals if separated from flock mates.[6,7] This behavior probably evolved in early ancestors as a survival strategy, just as it has in several other

FIGURE 9.1   (a) Sheep are grazers *(photo compliments of Susan Schoenian; used with permission[2,3])* and (b) goats are browsers *(photo compliments from Bridget and Gary Sebesta; used with permission).*

species. There is safety in numbers. Individuals are less exposed to predators, have multiple individuals to mount a defense, and in some species, the movements are confusing to predators as well. Solitary animals are more exposed to threats on all sides and must depend on their ability to fight, flee, or hide for survival. Lambs tend to stay close to their mothers, even in play. Goats are more individualistic and curious, so it is not uncommon for them to go their own way to check something out. They do live in groups, but the herd is less cohesive than is the sheep's flock. Kids are more active than lambs and are frequently seen jumping and climbing onto elevated objects (Figure 9.2).

The expression of aggressive behaviors by the males of both species occurs in different ways. Rams use head butting, in which they charge each other straight on until their upper foreheads collide. One or both goat bucks will flick their tongues, rear up on their hind legs, slightly rotate their necks, and then lunge until their heads collide. Play behaviors of the young mimic parts of these fighting behaviors. Mock fighting allows individuals a chance to practice techniques so that they become more proficient at the moves. It also increases muscle tone and physical strength needed to establish themselves within the hierarchy of the flock/herd. Head butting usually peaks during the prebreeding (rutting) season.[7]

FIGURE 9.2   Compared to sheep, goats are more independent of the flock. Both kids and adults are more active and are commonly seen climbing onto vertical objects. *Photo compliments of Bridget and Gary Sebesta; reproduced with permission.*

The amount of aggression is greatly reduced when estrus ends. The use of the heads in fighting for both species is a warning for people who work around sheep and goats. The most efficient way to reduce aggression in an animal or herd is to avoid precipitating it. Avoid petting the animals on their head, particularly the males. It can trigger the aggressive behaviors directed at the person who is doing the petting. Stepping back from the animal just before it behaves aggressively is another important way to reduce the potential for it to resort to aggression. Stepping back during aggression rewards the aggressive behavior; however, in life-threatening situations, the handler would seek safety and then avoid repeating the stimulus that resulted in the aggressive episode. Change the angle of approach, move the animal, or trap the animal before approaching. If aggression in animals works for them, it will likely be repeated.

Aggression is typically associated with dominance, although it does not have to be. Hierarchies are more evident in goat herds than in sheep flocks.[2] This is partially related to the presence of horns in goats. Historically, the reduction in horn growth allows energy to be diverted to other important functions, such as successful reproduction. When the two species are kept together, goats will usually dominate sheep because of their more aggressive nature, but if raised together, a ram usually is dominant because he can head butt the billy's abdomen while it is still rearing up.[6]

In several ways, goats are more sensitive to their surroundings than are sheep. While parasites are a problem for both species, goats do not seem to tolerate a heavy burden as well as sheep do, probably because their browsing pattern would normally protect them from heavy infestations but grazing would not. They also are more apt to sulk, hide in corners, or lie down in order to resist handling.

FIGURE 9.3   (a) Sheep flock together and tend to move and live in close proximity to each other. (b) In large pastures, they will spread out somewhat. *Photos compliments of Susan Schoenian; used with permission.*[6,7]

SHEEP BEHAVIOR

In the sheep industry, gathering the animals together is called flocking, banding, or mobbing. As a survival behavior, grouping has obviously been effective, and it has been selected as a desirable trait for domesticated sheep as well. Groups of animals that maintain close proximity to herd mates or are in a familiar environment seem to remain less reactive (Figure 9.3). Gathering sheep is often aided by the use of dogs, particularly in large pastures or on open rangeland. Because of their strong

FIGURE 9.4   Lambs learn to follow their mothers and then learn to follow other members of the flock because of their mother's behavior of staying close to flock members. *Photos compliments of Susan Schoenian; used with permission.*[6,7]

liking for food, a bucket of grain can make gathering easier in smaller pastures. The strong instinct of sheep to stay close together can have unintended consequences. For instance, in 2006, 400 sheep plunged to their death in Eastern Turkey after one of the sheep tried to cross a 15-meter deep ravine and the rest of the flock followed.[7]

Shortly after birth, lambs remain close to their dam. The lamb-following-mom behavior is both instinctive and learned. For several days after birth, lambs will follow any large moving object, so handlers must use some caution to be sure lambs learn to follow their mother consistently.[3] Lambs also remain close to the ewe during sleep, perhaps to initially help with thermoregulation.[5] The lamb learns to follow the flock by association with the ewe's behavior of following the flock (Figure 9.4). Because ewes tend to group, their lambs learn that too. Lambs reared with goats or other farm animal species will also exhibit grouping with the resident species.[7]

Separation is very stressful, as evidenced by bleating vocalizations by both the lamb and ewe. When reunited, the lamb will immediately nurse.[3] Ewes calling lambs when the flock is being grouped or moved can cause an association of stress vocalizations with the presence of humans. If instead the lamb hears the ewe's call during feeding behavior, the lamb can learn to seek food instead.

Wild species of sheep follow a strict order of who follows who when on well-defined paths. Domesticated sheep kept to multigenerational flocks are usually led by the oldest ewe, followed by her offspring, with each of them followed by their offspring. The flock leader is usually the oldest female with the most offspring. This pattern is broken up when sheep from different sources are mixed.[3] This suggests an age-related learned component.

Sensory capabilities for sheep are similar to those of other livestock species, as discussed in Chapter 2. Detailed vision was not particularly important for survival, but finding movement was. As a result, handlers often note that in unfamiliar pens, sheep do not easily find open gates. Once the animals learn the whereabouts of gates and have traveled down alleys, they will readily enter and exit alleys and chutes, assuming negative things have not happened there. Sheep avoid unfamiliar shadows, definitive contrasts between light and dark, and unfamiliar places. While sheep are not quite as sensitive to sound as other livestock species (Figure 2.6), they still can be startled with sudden, intense sounds. They will become reactive and more difficult to handle, so the elimination of intense noise is a must. Reactive sheep will typically show teeth grinding as a sign of distress. Relative to scents, sheep can learn what predators smell like; although there is also evidence that certain odors are instinctively recognized.[5,4] This may be the reason sheep tend to graze facing into the wind. Olfaction is the main method used by ewes to locate their lambs. Sheep also use smell to locate water and discern plants and feed.

Unfamiliar environments, intense or abrupt sounds, handlers approaching a sheep's blind area, new handlers emitting unfamiliar sounds, rough or abrupt touching and handling, or excitable humans tend to cause reactivity. This increases the likelihood that sheep will try to escape or avoid the source. Escape behavior can then be learned. This is complicated further in that the escape behavior appears to be infectious to the rest of the flock. Low-energy handling is as important for sheep as for other livestock. Animals that tend to escape at the sight of handlers should be contained and conditioned to accept handler presence. Teaching the sheep to herd from place to place in a pen or pasture is as important now as it was at the time of domestication.

GOAT BEHAVIOR

Goats are gregarious, but they are more independent than sheep. If a person walks between two sheep, they would appear distressed and try to get back together. Goats would go their own way. The strongest social groups within a herd are composed of related individuals. Goats that are leaders have more related individuals around them.[2]

Familiarizing goats individually or in groups to handler action, facility structure, environmental motion, and noise will go a long way toward avoiding problems. It reduces their tendency to sulk or escape and makes them less reactive. As with other livestock species, reducing external stimuli when needing to handle individual animals is important. This includes slowing handler and equipment motion, reducing unnecessary noise, modifying pain-associated movements, and employing efficient facility design and use. In general, calm humans have minimally reactive goats. Handling the animals in a humane, efficient manner for several minutes each day after feeding will generally accustom the animals to human presence and moving to human signals.

The senses of goats are similar to those of other livestock species. While the blind spot of sheep might be increased by wool growth on the cheeks, ear tags can block some of the peripheral field for goats.

HANDLING SHEEP AND GOATS

Sheep are the easiest to move as a group because of the strong cohesiveness of their social bond. They have "active followership." If one goes, they all go. As was mentioned at the start of this chapter, loading sheep can be quite easy if this strong bond is taken into account. Start by gently restraining one ewe in the front of a trailer. Her bleating attracts flockmates and then a handler can use the concepts of herding discussed here to move the rest up the ramp and onto the trailer. Similarly, a Judas goat trained to walk through the pen of sheep will lead them up the ramp. The goat will circle around inside the trailer and wait by the door to exit when the last ewe is loaded. A well-trained goat is a prized possession to any trucker who hauls a lot of sheep.

Sheep kept in large flocks avoid close proximity with people, and as soon as someone approaches the ewe's flight distance, she moves off (Figures 2.10 and 2.11). Controlled movements into and out of this flight distance are used to move sheep slowly. The entrance location and the direction of movement of the handler determines which way the animal goes (Figures 9.5 and 9.6).

Teaching all sheep or goats in the group to herd as a unit is best started in a large pen where they are comfortable. It is based on the flight distance and the social bonding of the animals. The handler should first enter the pen on the side opposite the feed bunk. The animals are encouraged to move toward the feed bunks by the handler walking left to right in a zigzag pattern across the pen (Figure 9.7). Once the majority of animals are in the area of the feeders, the handler stops and only rocks left to right as needed to keep the sheep in the feeder area. After a couple of minutes, the handler should slowly move along

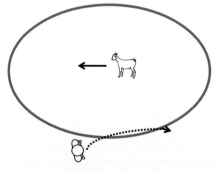

FIGURE 9.5   Slow, controlled movements into and out of the goat's flight distance are used to move the animal. The direction of movement of the handler determines which way the animal goes. © *Don Höglund (2015).*

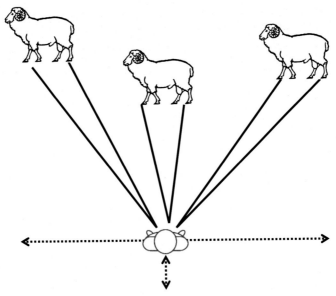

FIGURE 9.6   The starting position of the handler is different with respect to the three rams. If the handler moves toward the rams, the lead one moves forward, the middle ram moves right and potentially forward following the lead animal, and the trailing ram likely moves backward. If the handler moves parallel to the animal movement, it will cause the animals to move forward or to slow or to stop, respectively. © *Don Höglund (2015).*

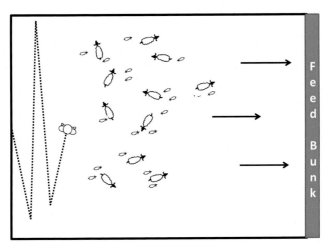

FIGURE 9.7   To teach the flock members to herd, the handler should first enter the pen on the side opposite the feed bunk. The animals are encouraged to move toward the feed by the handler walking left to right in a zigzag pattern across the pen. © *Don Höglund (2015).*

one wall of the pen, moving toward the sheep or goats on that end. This causes
the animals to move toward the others. The handler continues to move parallel
to the fence, but several feet from it. He or she should use a left-to-right rocking motion and a slow forward motion to move the animals to a new location
(Figure 9.8). The handler should hold the animals there for a couple of minutes,
and then repeat the process of moving to a new location and holding three times,
ending back at the feed bunk area.

When gathering sheep or goats after they have had the initial training, the
handler would enter the pen and move along the edges to a position opposite

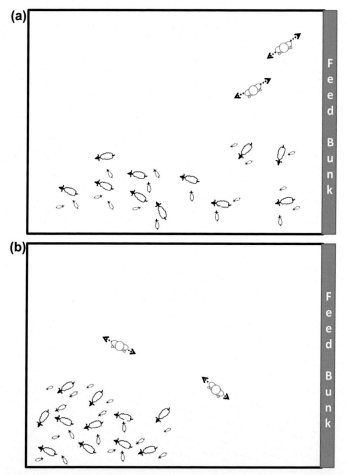

FIGURE 9.8 After a couple of minutes of being held near the feeders, (a) the handler should move
slowly along one wall and then the along the feed bunk of the pen. This causes the animals to move
toward the others. The handler continues to move parallel to the side of the pen, but several feet
from it, (b) using a left-to-right rocking motion and a slow forward motion to move the animals to
a new location. © *Don Höglund (2015)*.

from the intended goal. Then the handler would use the zigzag pattern across the width of the pen until the animals gather together and move forward. Each part of the zig or zag should go farther to the side than the location of the animal on that side (Figure 9.9).

Once trained, each generation teaches the next. The kids and lambs learn by following mom, making separate lessons unnecessary.

Goats, and occasionally sheep, may be more reactive and may keep running instead of walking. They can learn to slow down using negative reinforcement and "the wheel" technique. The basic concept is to move the animals along the

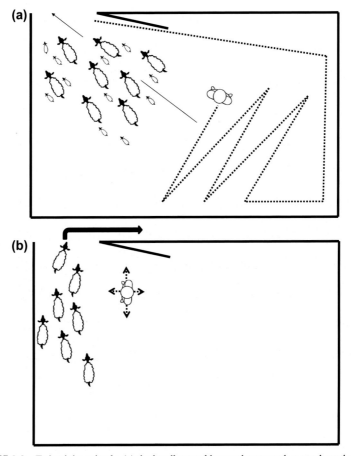

FIGURE 9.9   To herd the animals, (a) the handler would enter the pen and move along the edges to a position opposite from the intended goal. Then the handler would use the zigzag pattern across the width of the pen until the animals gather together and move forward. Each part of the zig or zag should go farther to the side than the location of the animal on that side. To cause the animals to turn in a particular direction, (b) the handler takes a position near the point where the curve is to happen and uses a rocking motion to encourage movement. The arc occurs because the animals are keeping visual contact with the person. © *Don Höglund (2015).*

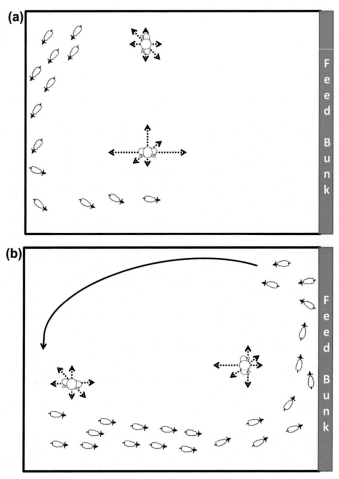

FIGURE 9.10   Reactive animals can be taught to slow down using the "wheel" technique. The basic concept is to (a) move the animals along the edges of the pen. As they start moving at a slower speed, the handler takes a small step back to reinforce the slower speed. The next step back will occur for an even slower speed. (b) The handler uses a variety of rocking motions to control the direction of movement so that the animals continue to circle the pen. © *Don Höglund (2015).*

edges of the pen. As they move at a slower speed, the handler takes a small step back to reinforce the slower speed. The next step back will occur for an even slower speed. The handler uses a left-to-right rocking motion, a forward-backward motion, and a directional motion that follows the animals (Figure 9.10).

Sheep are easier to work as a group, or at least with groups of approximately five animals. Goats can be worked alone or in groups. Moving an individual down an alleyway, along a fence, or in a pasture is done using a zigzag pattern. This technique uses an intrusion into the animal's flight distance, but the left-to-right crossing is done in such a way that the animal can see the handler so

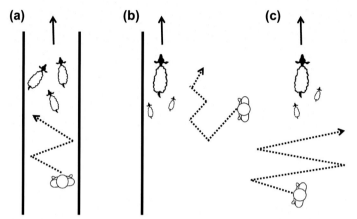

FIGURE 9.11   Moving one or more animals (a) down an alleyway, (b) along a fence, or (c) in a pasture is done using a zigzag pattern. This technique uses an intrusion into the animal's flight distance but in a way that the animal can maintain visual contact. Straight lines make the movement more predictable to the animals. © *Don Höglund (2015).*

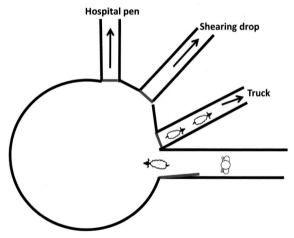

FIGURE 9.12   The sorting pie is a variation of the redirection pen, which is useful when animals must be frequently sorted to be weighted, dipped, or loaded. © *Don Höglund (2015).*

that movement continues straight forward (Figure 9.11). Straight lines make the handler's motion predictable, and the degree of angulation controls the speed of the animals.

The redirection pen and sorting pie can be used for sheep and goats just as for cattle and described in Chapter 6. They permit efficient and controlled movement of animals. The redirection pen is helpful for loading animals onto trailers and trucks. Large operations that frequently use scales, dipping vats, and trailers, or that sort animals frequently, find the sorting pie to be handy (Figure 9.12).

REFERENCES

1. Chance P. *First course in applied behavior analysis*. Long Grove (IL): Waveland Press; 2006. p. 485.
2. Houpt KA. *Domestic animal behavior for veterinarians and animal scientists*. 3rd ed. Ames (IA): Iowa State University Press; 1998. p. 495.
3. Hulet CV, Alexander G, Hafez ESE. The behavior of sheep. In: Hafez ESE, editor. *The behaviour of domestic animals*. 3rd ed. Baltimore: The Williams and Wilkens Company; 1975. p. 246–294.
4. LeDoux JE. Rethinking the emotional brain. *Neuron* February 23, 2012;**73**(4):653–76.
5. LeDoux JE. Coming to terms with fear. *Proc Natl Acad Sci* February 25, 2014;**111**(8):2871–8.
6. Schoenian S. Sheep 101, http://www.sheep101.info/index.html.
7. Schoenian S. Sheep 201, http://www.sheep101.info/201/index.html.

Index

Printed in the United States
By Bookmasters